D1539017

Talking Climate

Adam Corner • Jamie Clarke

Talking Climate

From Research to Practice in Public Engagement

Adam Corner
Climate Outreach
Oxford, United Kingdom

Jamie Clarke
Climate Outreach
Oxford, United Kingdom

ISBN 978-3-319-46743-6   ISBN 978-3-319-46744-3 (eBook)
DOI 10.1007/978-3-319-46744-3

Library of Congress Control Number: 2016952792

Cover illustration: Pattern adapted from an Indian cotton print produced in the 19th century

Printed on acid-free paper

This Palgrave Macmillan imprint is published by Springer Nature
The registered company is Springer International Publishing AG
The registered company address is: Gewerbestrasse 11, 6330 Cham, Switzerland

PREFACE

The question of how to communicate about climate change, and build public engagement in high-consuming, carbon-intensive Western nations, has occupied researchers, practitioners, policy-makers, campaigners, and community organisers for more than two decades.

During this time, and mirroring the glacial pace at which international political negotiations have progressed, limited progress has been made. Socially and culturally, climate change remains (for the most part) the preserve of a committed band of activists. The 'carbon footprints' of many Western countries – and the citizens of these nations – remain high. The public conversation about the energy system is mainly focused on the costs of household energy bills. Public engagement is stuck in second gear.

Meanwhile, the predictions made by scientists over the past quarter of a century are beginning to come true. Pick any metric – levels of atmospheric CO_2, global average temperatures, ocean acidification, the prevalence and severity of certain types of extreme weather – and it is clear that the climate is changing, with all the risks and dangers this entails for human societies and the natural world. As countless analyses, declarations, and calls to arms have made clear, rapid and radical changes to the social, economic, agricultural, transport, and energy systems of the world are required if the ambition of the UN agreement reached in Paris, in 2015 (to limit levels of global warming to less than 2 degrees centigrade above pre-industrial levels), is to become a reality. A widespread and sustained shift in public consciousness and engagement is a central part of this challenge.

The purpose of this book is to outline how public engagement with climate change can shift *out* of second gear. There is no lack of relevant academic evidence, but most of it is not connected with the practitioners who can put it to good use. Similarly, campaigners on a whole range of issues have developed a huge amount of learning (often through trial and error), but these lessons are not consistently applied, and climate change continues to remain trapped in the 'green ghetto'.

In this book we offer a practically oriented evidence base for why (and how) practitioners could do things differently. The pieces of the puzzle already exist to make this happen – in academic research and practitioner expertise – but a coherent new agenda for public engagement is required to make these pieces fit together. The five principles outlined in this book offer a fresh approach to a familiar problem. By spanning the full width of the space between primary academic research and applied practitioner strategies, we hope the book will be relevant for academics, educators, campaigners, and communicators.

This book does not contain a prescriptive set of rules or a 'how to' guide for running the perfect climate change or energy campaign. What it offers is a set of principles, all grounded in academic evidence, that together form a fresh approach to public engagement, providing a platform on which individual initiatives and campaigns can be built:

> *Principle 1: Learn lessons from previous campaigns, and be prepared to test assumptions.*
> *Principle 2: Public engagement should start from the 'values-up' not from the 'numbers-down'.*
> *Principle 3: Tell new stories to shift climate change from a scientific to a social reality.*
> *Principle 4: Shift from 'nudge' to 'think' to build climate citizenship.*
> *Principle 5: Promote new voices to reach beyond the usual suspects.*

These five principles are described in detail in subsequent chapters, alongside a proposal for a fresh approach to widen and deepen public engagement, as well as suggestions for how we would know that the approach was 'working'.

Our societal response to climate change is not something that will be 'solved' in a generation: the question of how we should live in a climate-changed world is one that will be relevant and essential for centuries to come. So although time is clearly of the essence, effective public engagement is not something that can happen overnight or by focusing on 'quick wins' at the expense of a more holistic understanding of the challenge.

But a robust foundation of public engagement and dialogue can ensure something more important than quick wins: a level of 'climate citizenship' that locks in the stuttering technological, economic, and political progress where the 'big wins' are to be found. From the uptake of energy-saving technologies, to the mandate offered to national leaders, to the social momentum behind new initiatives like fossil fuel 'divestment', public engagement underpins it all.

It follows that we should invest in public engagement on climate change in the same way that we invest in every other dimension of the challenge of decarbonisation. This book offers the principles that can catalyse public engagement and the promise of shifting the climate change discourse out of the margins and into the mainstream. It is time to start talking climate.

Acknowledgements

The perspectives and arguments contained in this book are the culmination of many years of thinking, research, discussion, and debate with friends and colleagues on the question of how to widen and deepen public engagement with climate change. Many of the ideas discussed in detail here have surfaced in previous publications, reports, and blog posts. We would like to thank Ewan Bennie (at the time a member of the Energy and Community team at the Department of Energy and Climate Change – DECC) for his foresight in commissioning a review of the literature relevant to building committed public support for energy system transformation, where several key concepts in this book were first explored.

All of our colleagues at Climate Outreach have played an important role in shaping our views, and we would like to thank them – and the Climate Outreach Board of Trustees – for their support for us while writing this book. We would like to thank in particular Dr Chris Shaw and George Marshall for their intellectual input, both in terms of ideas sharpened in conversations and practical advice on how to present the arguments in the book clearly and concisely. Anna MacPhail provided invaluable support in sourcing and standardising references and citations, and Clara dos Santos played an important role by supporting on background research for Chap. 2.

Dr Stuart Capstick (Cardiff University) and our Climate Outreach trustee Dr Jonathan Rowson (Perspectiva) both provided extremely helpful reviews of the book in advance of its completion, and we are very grateful to both of them for their time and careful, considered reflections. Luke Sinnick, Professor Nick Pidgeon, and another Climate Outreach trustee, Adam Ramsey, provided additional comments and feedback on specific chapters. At the book-proposal stage, we were very grateful for constructive and thorough reviews from Professor Stephan Lewandowsky and an anonymous reviewer.

Contents

LIST OF FIGURES

LIST OF TABLES

CHAPTER 1

A Fresh Approach to Public Engagement

Abstract Chapter 1 introduces the core theme of the book: that to make meaningful progress on climate change, a fresh approach to public engagement based on five key principles is required. To date, public engagement initiatives have produced limited results: most people hold no strong views about climate change, a social silence is pervasive, and climate change has become a politically polarised issue in many Western nations. Even those who accept the reality of the problem are often dismissive of the role they can play in solving it – most people have not yet heard a story about climate change that sounds like it was written 'for them'. In this book we argue that a fresh approach is required, based on peer-led participatory engagement – talking climate – not 'nudging' people into small changes in energy behaviours.

Keywords Public engagement · Fresh approach · Communication · Participatory · Principles · Climate change · Talking climate

1.1 The Challenge of Communicating Climate Change

Communicating climate change is not easy. Despite more than two decades of awareness raising and campaigns, and a sprawling academic literature on the subject, public engagement remains (for the most part) stubbornly stuck in 'second gear'.

© The Author(s) 2017
A. Corner, J. Clarke, *Talking Climate*,
DOI 10.1007/978-3-319-46744-3_1

1

Perhaps this should not be surprising: a frustrating inertia has defined almost every aspect of our response to climate change, from the moment that James Hansen's testimony to the US Senate catapulted the idea of a rapidly warming world out of the scientific literature and into the political and public discourse. As dozens of analyses can attest (Hulme 2009), climate change is a complex, multifaceted problem, seemingly unlike anything we have ever consciously faced before. The range of analogies and metaphors that are mobilised to describe and deconstruct climate change (Nerlich et al. 2010; Ereaut and Segnit 2006) speak volumes about the difficulty of finding a suitable point of comparison. Should we declare 'war' on carbon? Is our planet 'sick'? Do we need a 'Marshall Plan' to rebuild our energy system analogous to the effort to rebuild after the Second World War?

The causes and the consequences of climate change are spatially and temporally displaced, providing an awkward fit with our moral sense-making machinery and confounding simple explanations of who is to blame for a changing climate (Marshall 2014a). Making sense of climate change requires expertise and understanding that cuts across disciplines, sectors, and cultures: there is no facet of human experience that is not related in some way to the climate in which we all live. And as a result, there is no simple way to disentangle our response to the risks of a changing climate from the kaleidoscope of social, political, and economic factors that govern our lives (Grundmann 2016). When we talk about climate change, what we are really talking about is ourselves, the things we value, and how we want the future to be (Hulme 2009).

Part of the reason that public engagement on this issue is so difficult is that comparing climate change to other 'issues' and social challenges facing society (like promoting healthy eating, or preventing the spread of HIV) is fraught with difficulties: in almost every instance, climate change seems somehow 'different'. As we argue in Chap. 2, this does not mean that we cannot learn valuable and important lessons from campaigns on other societal challenges. But climate change is not a 'single issue' or even a suite of related issues: it is a lens through which we can examine almost every aspect of our lives, from the global and political, right down to the personal and everyday (Hulme 2009; Grundmann 2016).

A central challenge for climate change communicators is that for most people, most of the time, there is no tangible, concrete 'signal' that climate change is taking place. The only relevant evidence of climate change that most people can directly experience for themselves is the local weather and seasons they encounter. But because climate change is a global and distributed problem – with rising global average temperatures producing a cluster of complex weather impacts in different regions – local weather is often not a good guide to the global climate. For citizens of nations where climatic conditions are generally moderate, climate change is too easy to dismiss as something that happens to other people, in other places (Weber 2010). And even when extreme weather events do occur, it is difficult to make simple statements about their relationship to underlying changes in the climate (Marshall 2014b). 'Encountering' climate change via extreme weather is certainly no guarantee that people will engage with the issue more generally (Reser et al. 2014).

And the complexity of climate change is only half the challenge. Communicating the myriad of ways in which our use of energy contributes to climate change (through travel, heating our homes, industrial processes, and our agricultural system) is also fraught with difficulties. Just like the climate system, the energy system is distributed, relatively abstract, and invisible to most people, most of the time. Energy is embodied in a range of behaviours and social practices that span the full spectrum of human activity (Henwood et al. 2015), and so altering these (or even making them conscious and visible in the first place) is an enormous challenge. Communicating climate change quickly merges into engagement with transforming the energy system and other potential responses to climate change – with an almost endless number of different permutations to consider.

It is all too easy to get lost in the idea of climate change or boggle at the all-consuming nature of the challenge of transforming our energy systems. But despite the complexity, we face very tangible risks as our climate changes, and profound choices about how we should live in a carbon-constrained world. Through the cumulative impact of human activities over predominantly the past 150 years, we have initiated an ongoing climatic experiment with an outcome that we cannot confidently predict. Significant impacts attributable to human influence have already begun (Fischer and Knutti 2015), and climate change is now 'with us' for the foreseeable future.

Whatever its limitations, the agreement reached at the UN negotiations in Paris during December 2015 (assuming the rhetoric of the agreement is matched by decarbonisation plans in reality) is a signal: the decision to take climate change seriously has now finally, formally been made. But even the most optimistic observer would not dispute that we have reached the starting line about 20 years too late, or that the vast majority of the hard work remains ahead of us. We continue to struggle to summon the collective will power to rise to the challenge that is already shaping the contours of the twenty-first century through issues like food insecurity, incidences of extreme weather, ocean acidification, and the relentless onwards march of global temperatures (Kelly et al. 2015; Nagelkerken and Connell 2015).

The question of why the world has not been quicker to react to climate change could be answered in a dozen plausible ways. The politics of power, a ubiquitous and entrenched extractive energy indus-try powering the global economy, geopolitical disputes over territory and energy security, and obstinate national self-interest have all played a role. But central among the explanations for our inertia in the face of climate change is our own psychology – that old chestnut 'the human condition' (Pidgeon 2012). In many ways we are not 'wired' to deal with a challenge like climate change (Marshall 2014a; Stoknes 2015; Moser 2016), and so responding to it in a meaningful way is a perpe-tual uphill struggle.

Given all of this, wouldn't it be easier to simply skip the public-engagement part and let the engineers and policy-makers get on with the serious business of solving climate change on our behalf? Is there time to spend on the slow, messy, laborious process of public engagement?

1.2   WHY PUBLIC ENGAGEMENT WITH CLIMATE CHANGE MATTERS

It is reassuring to draw sharp distinctions between politics, technology, and individual-level attitudes and behaviour. And it is true that the actions of any one individual are inconsequential in terms of reducing human influence on the climate. A very literal (and atomistic) reading of how the beliefs and behaviours of individual citizens relate to climate change would suggest that building public engagement with energy and climate change is not an impor-tant priority. But the reality is that public attitudes and social practices almost always play a central role in political and technological decisions (Whitmarsh et al. 2011; Spurling et al. 2013). Governments in democratic nations will not run significantly 'ahead' of where they perceive public opinion to be (Pidgeon 2012). The most carefully considered policy interventions will backfire if they don't take account of how people will respond. Seemingly 'win-win' technologies and ideas (such as free home insulation schemes) will not be taken up if they are unpopular or viewed as irrelevant. Hard-headed corporate decision making may seem immune to the complexities of human emotions and social norms, but it is not (Wright 2016). Decisions about how to adapt and adjust to a changing climate are fundamentally linked to the way in which people understand the problem of climate change (and who they see as having the responsibility for dealing with it).

Of course, policy change is *possible* without public support, and policies can drive behavioural changes and industrial practices (so the relationship is certainly not all one-way traffic). But as we argue throughout this book, a proportionate societal response to climate change is not something that can be achieved unthinkingly, or simply via cracks of the regulatory whip. What ordinary people think about climate change – and their perception of what it means for their lives – really matters. Our collective decision making is the beginning, the middle, and the end of the climate change story.

Public engagement is the foundation on which technological, political, and economic policies on energy and climate change are predicated (Parkhill et al. 2013). Only when – at a societal level – the importance of the issue has been collectively internalised, can we expect climate policies to stand robustly against potentially hostile political winds. Legally binding targets for emissions reductions are crucial, but represent only one-half of the deal; informed public support for achieving them is the other. Our argument is that the stories that we tell ourselves about what climate change means, who is responsible for responding to it, and what this response should look like are just as important as the technologies, laws, and policies that will usher in a more sustainable world (Smith et al. 2014). But so far, we have not been very successful at catalysing progress on this front.

While it would be an overstatement to say that public engagement with climate change is absent altogether, the public discourse is characterised by a 'wide but shallow' sense of the seriousness of the challenge at hand. Although in some nations (particularly in the Global South, where per capita emissions are low and vulnerability to climate impacts is high), there are extremely high levels of concern about the risks of climate change, in many others (particularly in wealthier, industrialised nations) climate change is a relatively low priority, even if awareness of the issue is high (Stokes et al. 2015). And in these wealthy, high-consuming countries (such as the US, the UK, and Australia), public engagement is characterised by attitudinal polarisation driven by ideological and values-based differences and a lack of meaningful engagement with the sorts of societal changes that a proportional response to climate change will entail (Hornsey and Fielding 2016).

1.3   TALKING CLIMATE?

Surveys in the US, the UK, and Australia show that when prompted with a survey question, large numbers of people are at least somewhat concerned about the risks posed by a changing climate and broadly supportive of a transition to a low-carbon society (Pidgeon 2012; Reser et al. 2014; Leiserowitz et al. 2011). But climate change is rarely listed among the most pressing issues facing individual countries or for individual citizens (even if at a global level it is more widely recognised – Pew Research Centre 2015). Most people hold no strong views about climate change: even those who accept the reality of the problem are often dismissive of the role they can play in solving it or quite literally do not spend any time thinking about it or talking about it (Rowson 2013; Leombruni 2015).

Some have described this as climate 'fatigue'. But that would suggest that people were sufficiently engaged with the issue in the first place to have become tired by it. Others suggest that public 'disavowal' of the problem is a reaction to what they consider to be the terrifying psychological reality of facing climate change (Weintrobe 2012). But have most people noticed that the consequences of unchecked climate change are so significant that they must bury their emotional reactions to it under a bundle of psychological defence mechanisms? Another popular explanation of the relative unimportance of climate change in the public mind is that people have a 'finite pool of worry' (Weber 2010), and that other, more immediately pressing issues such as economic problems or the risks of terrorism take precedence in the concern hierarchy. Certainly, there is evidence that concern about climate change has ebbed and flowed along with macro-scale social and economic problems (Brulle et al. 2012), and many problems (food insecurity or political instability) are perceived as more pressing than climate change. The problem, of course, is that climate change increasingly plays a role in all of these other issues, often exacerbating them.

A full 40 % of 2000 British people surveyed in 2013 (Rowson 2013) said that they never speak about climate change to their friends, family, or colleagues. The few conversations that were held tended to be short, with more than two-thirds of people talking for less than ten minutes about the issue. Another study a year later found similar patterns: 40 % of survey respondents said they discussed climate change with family and friends at least sometimes, though for most people climate change was not something that arose in discussion particularly often. Twenty per cent said they never discussed the subject with family and friends (Capstick et al. 2015a). Statistics from Yale University paint a similar picture of US public opinion. A vanishingly small proportion of Americans were talking about climate change to each other in 2015: only 19 % of the US public hear about climate change in the media more than once a week (with only 4 % talking about climate change with others once a week – Leiserowitz, et al. 2015).

Moreover, a range of indicators suggest that levels of concern (as captured in surveys) do not straightforwardly translate into more meaningful behavioural outcomes or even higher support for low-carbon policies (Capstick et al. 2015b), and there are some serious question about how 'deep' public engagement with energy and climate change is. For example, a report by the British Royal Society of Arts used national survey findings to identify a large majority of the population who could be described as in 'stealth denial' about climate change (Rowson 2013). That is, while many people expressed high levels of concern about the issue, a majority also agreed that there was little they personally could do to contribute to tackling climate change or said that they did not feel uneasy about climate change.

As the work of anthropologist Kari Norgaard has so vividly demonstrated, people are capable of a spectacular form of doublethink – socially constructed silence – when necessary. Over the course of 2 years, Norgaard interviewed 46 people in a remote coastal town in Norway. Awareness of climate change was high and people openly recognised that the weather was changing dramatically. In particular the ski hill, an essential component of the town's local economy and cultural identity, was opening weeks later and only with the help of artificial snow. Despite this, there was virtually no discussion about climate change. As a local teacher put it to her 'We live in one way and we think in another. We learn to think in parallel. It's a skill, an art of living' (Norgaard 2011).

It is certainly easier to not think or talk about climate change. Even those who are concerned about the problem may feel deeply conflicted about acting on it. We have more than enough psychological tricks up our sleeves to procrastinate the problem away. So we must get better at overcoming these self-inflicted barriers and obstacles and derive a theory of public engagement that recognises that it is 'us' (i.e. people) not 'the climate' that should be the main focus of attention. A fresh approach to public engagement is required.

1.4   A Fresh Approach: Five Principles for Public Engagement

The evidence base we draw on in this book is derived almost exclusively from Anglophone nations, and so as a result, the primary focus of this book is on building public engagement in these high-emitting nations, and their non-Anglophone counterparts in the wealthy, Western world. We do not dismiss the importance of building public engagement in developing nations, but the context is very different: many developing nations will increase per capita energy use in the immediate future. So while understanding climate change in the minds of (for example) Bangladeshi or Ghanaian citizens is critical, the principles for public engagement we outline in this book are primarily focused on the citizens of high-emitting nations.

As well as the growing research base on this issue, there is (as we discuss in detail in Chap. 2) a long history of campaigns to engage the public on energy and climate change. These campaigns have tended to operate relatively independently of the academic literature, and campaign materials are rarely tested or evaluated. As a result, a significant amount of academic research has been dedicated to analysing – and in some cases challenging – activists' strategies. So while there are a raft of resources available to communicators, a long track record of public-facing campaigns on climate change, and an academic evidence base that is growing by the week, the intersection between them is limited. Building much stronger bridges between research and practice on climate change communication is a key element in developing a new approach to public engagement – something which this book seeks to address.

For too long, public engagement on climate change has been dominated by well-intentioned but short-term, or simply ill-considered strategies of communication. Very few campaigns or initiatives have been able to communicate about climate change in a way that effectively speaks to a broad range of public values: for most people, climate change is a scientific but not yet a 'social reality'. Most people have not yet heard a story about climate change that sounds like it was written for them, in language that connects with their interests, values, or identity. Can the deadlock in public engagement be broken? Our argument in this book is that in order to break it, a fresh approach is required (Corner and Groves 2014).

The remainder of this book describes five principles for public engagement on climate change. It joins the dots between the academic evidence base, and campaign strategies. In Chap. 2, we consider what lessons can be drawn from public engagement on other significant societal issues, and how these lessons are being applied in energy and climate change campaigns (**Principle 1**). The central message of this chapter is that there are important lessons to be learned, but that in many ways climate change does not easily fit the mould – and that the careful testing of assumptions and campaign materials is therefore crucial.

Next (**Principle 2**), we ask whether climate change has been approached from the wrong perspective and argue for the importance of working 'upwards' from people's values and worldviews (Corner et al. 2014), rather than 'downwards' from the big numbers that have characterised so many previous campaigns (Chap. 3). In this chapter, we make the case for focusing on 'communal' rather than 'self-focused' values as the basis for public engagement (Crompton 2010). People from across the political spectrum identify with different communal values (Corner 2013a), and these provide the building blocks for building engagement with energy and climate change.

Most people deal in stories and anecdotes, not graphs and statistics, and in Chap. 4 we review the growing literature – from across the social and political sciences – on the importance of new 'narratives' on climate change that resonate more effectively with a diverse range of public values. **Principle 3** centres on the notion that stories – rather than scientific facts – are the vehicles with which to build public engagement (Smith et al. 2014).

In Chap. 5, we consider the role of individual behavioural changes in a new approach to public engagement. Early campaigns to engage the public focused on the 'simple and painless' changes in behaviours (such as switching off lights) that it was hoped would lead to more significant lifestyle changes. More recently, social marketing-based strategies like the 'nudge' approach have attracted a lot of interest (Sunstein and Thaler 2009). But sustained and substantive changes in the behaviours of individual citizens have not been forthcoming (Capstick et al. 2015b). And although people can certainly be 'nudged' into certain limited pro-environmental behaviours, this does not involve them reflecting on why these changes matter (or how it might relate to other aspects of their lives – Evans et al. 2013). Chap. 5 sets out why it is crucial to move from 'nudge' to 'think' as a strategy for public engagement (**Principle 4)** and argues that participatory dialogues and conversations offer the best method with which to build a sense of climate citizenship. Individual behaviours matter, but only as part of a more integrated and holistic approach, where personal actions have a clear relationship to the bigger picture on energy and climate change.

In Chap. 6, we focus on the importance of promoting and amplifying 'new voices' on energy and climate change, diversifying the perspectives that define the climate change discourse, to help overcome the social silence that so often surrounds the issue (Corner 2013b). A broader social support base can engage groups of people beyond the 'usual suspects', moving climate change from a scientific to a social reality and positioning a proportionate response to climate change as something that is important to them (**Principle 5**).

The final chapter (Chap. 7) brings all five principles together to form a new approach to public engagement. Through peer-led participatory engagement – talking climate in existing social networks, and on an ambitious scale – the importance of climate change, and sense of ownership over building a societal response to it, can be dramatically shifted. Committed public engagement will not manifest itself in a generic prescription for a 'green lifestyle' or stand and fall on favourability towards a particular type of energy technology. Even among a population deeply supportive of the importance of climate change and accepting of their role in responding to it, there will be disputes, competing priorities, and differences of opinion.

We argue that the aim of public engagement should not be to quash or eliminate these differences, in the name of decarbonisation. Instead, the aim should be to encourage, promote, and facilitate public dialogue about climate change – but dialogue guided by reflection on the reality of life within the parameters of existing (in many cases now legally binding) targets for carbon reduction. The implication of viewing public engagement in this way is that there is no particular set of behaviours or one particular political viewpoint that has a monopoly on 'solving' climate change. To imagine that differences in political opinion or fundamental disagreements about values and worldviews could be smoothed over by climate change is naive and ultimately counterproductive.

With levels of atmospheric CO_2 at levels unprecedented in human history (and rising), the idea that climate change can be 'solved' in the conventional sense of the word is itself simplistic and unhelpful. This is a challenge that is with us, now, and which we will collectively have to manage for a very long time. Managing climate change is a journey, and the *process* of public engagement is as important as the specific changes in attitudes, behaviours, social practices, and policy preferences that it produces (although these tangible changes are also crucial and urgently required). Climate change is happening too fast for our psychological, social, and cultural capacities to adapt, and in our understandable collective impatience to summon an urgent societal response to this most time-sensitive of problems, the value of 'talking' can seem limited, when 'action' is what is ultimately required. But it is precisely this type of approach – climate conversations – that will ensure an urgent societal response is also a robust one. Hitting the fast forward button by fixing our sights on political goals and emissions targets (without enough focus on building the broad-based support required to achieve them) is a false economy.

Of course, there is a need for a social consensus that climate change matters. And committed public support for a range of far-reaching changes is crucial. But there will always be social, political, and cultural disagreement over exactly what these changes should be, and as the evidence in this book shows, this is not a reason for despair. The bewildering complexity of climate change may make it an almost-intractable problem to 'fix', but it also offers a mirror in which almost limitless possible futures can be glimpsed. By diversifying and deepening public engagement with climate change, distinct but deeply committed perspectives can be nurtured – the solid foundations that a proportionate societal response to climate change requires.

There will never be a point where a vibrant and dynamic public conversation about climate change is not a good idea. Public dialogue – among friends and peers meeting in person, or interest groups online – is the foundation upon which a sustained and coherent societal response to climate change can be built. But at present, this dialogue is almost entirely absent, and as a result, engagement with energy and climate change remains shallow, fragile, and superficial. As Susan Moser (a veteran and influential academic voice on climate change communication) argues in a recent paper reviewing the field, recommendations on public engagement can often seem insular or lack the perspective proportionate to the scale of the challenge, and there has been surprisingly little work focused on the idea of public engagement as an essential and valuable *process* in and of itself, rather than a means to an end:

> Despite a clear recognition in the scientific community how far-reaching and long-lasting human-driven climate change will be, despite the emerging literature on communicating climate adaptation, and the observation—and sometimes lament—of climate change issue fatigue, few have begun to grapple seriously with what it means to communicate, deal with, and engage publics around an issue—practically—forever . . . (C)limate communication practice and research must grapple with the question of what communication for the very long-haul entails, and what its function might be. (Moser 2016)

Our argument is that 'communication in the long term' can only be realised through ongoing participatory dialogue, providing the momentum and legitimacy for the far-reaching societal changes that will be required. Climate policies do not (and cannot) exist independently of public engagement and debate about the societal implications of climate change. As Amanda Machin argues in her book *Negotiating Climate Change*, there is no overarching grand green scheme that suits everyone, and differences in opinion should not be smoothed over (Machin 2013). It may seem counter-intuitive to promote dispute and disagreement – and of course, time is of the essence – but passionate disagreement in the short term is actually an essential prerequisite for substantive agreement in the long term. Without this foundation – in the absence of meaningful public participation – climate policies are a house of cards.

A vibrant and dynamic public discourse on climate change will inevitably mean negotiating disagreements, but this is preferable to the absence of a public discourse altogether. And campaigns in support of particular policies will only be effective if they are grounded in wider and deeper public engagement.

So this book is not a 'how to' guide for winning a campaign against a fracking well or increasing the percentage of people who cycle to work.

It does not contain a prescription for a green lifestyle or a technical specification for reducing emissions from aviation and shipping. But the approach we describe – five evidence-based principles for public engagement – provides the starting point for all of these campaigns and more. Only by starting with the fundamental factors that determine public engagement with climate change – beginning with people, rather than climate science – can a proportionate societal response be sustained.

–

CHAPTER 2

Is Climate Change Different?

Abstract Climate change developed as an 'environmentalist' issue in the public mind, but it is not straightforwardly comparable to other environmental issues. Major campaigns on poverty reduction, HIV/AIDs, and immunisation provide important lessons for public engagement, including the importance of focusing on 'movements' and interpersonal engagement, rather than 'moments' and mass communication. Radical activism has an important role to play if it complements the goals of mainstream public engagement, but too much focus on demonising 'enemies' may be counterproductive. The first of the five core principles for public engagement is *learn lessons from previous campaigns and be prepared to test assumptions.*

Keywords Campaigns · Lessons · Poverty · Environmentalism · Health · Testing assumptions

© The Author(s) 2017
A. Corner, J. Clarke, *Talking Climate*,
DOI 10.1007/978-3-319-46744-3_2

PRINCIPLE 1: LEARN LESSONS FROM PREVIOUS CAMPAIGNS AND BE PREPARED TO TEST ASSUMPTIONS

2.1   A SHORT HISTORY OF HOW CLIMATE CHANGE BECAME AN 'ENVIRONMENTAL' ISSUE

Climate change emerged on the public radar in the 1980s at a time when a superficially similar global emissions issue, ozone depletion, was making headlines. It also followed widespread engagement in the UK, the US, and Canada with another 'environmental' problem: acid rain, which had been prominent in the public mind since the 1970s (Menz and Seip 2004; Weiss 2012). The US and Canadian legislators adopted a 'cap and trade' system to deal with acid rain, reducing levels of sulphur dioxide (its main cause) and creating a market for firms to buy and sell government-issued allowances to emit the gas (Stavins et al. 2012). Once established, it was a policy solution that required no ongoing public involvement. A similar system was adopted by the 1987 Montreal Protocol to deal with ozone depletion: chlorofluorocarbons (CFCs) were rapidly regulated out of the industrial system, and the 'hole' in the ozone layer began to shrink.[1] Binding international law was adopted to enforce a market-based system of emission permits in what became viewed as 'the most successful environmental protection agreement ever reached' (Marshall 2014a).

Although they sparked dispute and debate at the time, both of these contemporary environmental issues were successfully addressed, and so it is unsurprising that campaigners and policy-makers in the field of climate change drew on the readily available precedents, made assumptions based on simplistic and metaphorical similarities, and assumed that the public would do the same (Marshall 2014a). So similar were the narratives between ozone depletion and global warming that for a long time, the public was thoroughly confused about them. Surveys as late as 1999 found that a quarter of Americans thought ozone depletion was the main cause of global warming (Lorenzoni and Pidgeon 2006).

All of this meant that there was a strong push towards characterising climate change as a primarily 'environmental' issue, to be dealt with via emissions caps, regulations, and the minimum of public engagement (this was an environmental problem, not a human one, after all). But running parallel to this trend was another formative influence on how climate change would come to be perceived: the wider notion of 'environmentalism'.

As environmental consciousness grew during the 1960s, 1970s, and 1980s (and organisations such as Greenpeace and Friends of the Earth, now considered 'household names', began to establish a real presence), climate change found a natural niche to evolve in (Dauvergne and Lebaron 2014; Hulme 2009). Understandably, for an issue with such strong roots in the natural sciences, early advocates of the cause were largely environmental pressure groups, concerned with the preservation of natural ecosystems and species. Because it was presented and discussed as primarily an environmental issue, it was not generally considered an economic, health, or social rights problem. 'Save the planet' was the dominant rallying cry, not 'power to the people' or 'what will this mean for the way we live our lives?'.

To be clear, significant progress has been achieved in the last 50 years by environmental pressure groups focused on addressing degradation from human impacts, whether in relation to pollution, resource depletion, or habitat destruction. As a force for change, the membership, and tactics and influence of the environmental movement have achieved a great deal, and their members are among the most committed to climate policies (Tindall and Piggot 2015). Environmental NGOs – and the many active grassroots organisations concerned with issue like ecological destruction, food sovereignty, or localised pollution – played an essential role in maintaining pressure on politicians and policy-makers during the early days of climate change campaigning. And it is just as well they did: many human rights NGOs were extremely slow to see the profound relevance of climate change for their work (Marshall 2014a).

But the strength of the environmentalist voice in the public discourse on climate change has had a lasting impact on the way it is represented in society today, and the cultural tropes of environmentalism have weighed heavily on the shoulders of climate change campaigners. The values and language associated with climate change are those of the 'environmentalist' – a stereotype which unfortunately does not have positive connotations for many people, to the extent that it can create resistance to the social change being advocated (Bashir et al. 2013). In our own work at Climate Outreach, we have repeatedly encountered strong, sometimes venomous, opposition to the iconography and language of environmentalism (Corner et al. 2015). And this pigeonholing of climate change in the public mind has had a lasting effect: despite being an issue with consequences for people's lives and livelihoods from across the social and political spectrum, climate change is not something that is central to most people's identity (and has become actively politically polarising in some nations). This is at least in part due to the close coupling between climate change and environmentalism.

When people are asked in surveys (Leiserowitz 2006) or in focus groups (Corner et al. 2015) what image comes to mind when they hear the term 'climate change', there is one overwhelmingly popular answer: polar bears. Iconic mega flora and fauna (from whales to pandas and rainforests) are a stalwart of large-scale environmental campaigns, so it is no surprise that a decision made by activists in the 1980s and 1990s to associate climate change with one iconic animal has provided a simple visual shorthand for the issue. But it has also reinforced the impression that climate change is a distant problem and arguably 'closed down' the climate discourse around a concept that is remote from people's day-to-day lives (Manzo 2010; Doyle 2011).

The 'ghettoisation' of climate change as an environmental issue might not have been problematic if it were more similar in nature to the problems of acid rain and ozone layer depletion that preceded it: addressing these issues required relatively minor shifts to the status quo, no major economic, social, and political upheaval, and next to no ongoing public support. Within this context, it is not surprising that the same type of approach was adopted by the UN when it came to establishing a model for climate change, and the same engagement processes were used to mobilise public support.

But the challenge of decarbonising the energy system in response to climate change is radically different. As is painfully clear in retrospect, the optimism created by the rapid and relatively straightforward success of the Montreal Protocol has not been mirrored in international negotiations on climate change. Twenty-one long years after they were initiated, the UN negotiations finally delivered a global agreement in Paris in 2015. The glacial pace of the negotiations is not the only one way in which climate change is different. While the primary pollutant implicated in ozone depletion – CFCs – was limited to a fairly small number of industrial product lines (and could easily be replaced with an alternative with the minimum of public disruption), carbon emissions are implicated in almost every aspect of our lives. Energy use is intertwined with our social practices in an intricate way (Henwood et al. 2015).

Our argument is that because deceptively simple similarities were perceived between climate change and the other 'environmental' issues of the time when climate change first emerged on the public radar, the wrong approach was often taken. Centralised and technocratic governance was favoured over strategies based on building, widening, and deepening public engagement, on the assumption that ongoing efforts to maintain public support and interest would not be necessary. And at the same time, the wider environmental movement – for all its positive gains and victories – subsumed the notion of climate change, stifling its identity in most people's minds and confining it to the 'green ghetto'.

2.2   The Coming Climate Catastrophe?

As iconic images such as a burning globe held in the palm of a human hand attest, the concept of catastrophe played a central role in many early climate change campaigns (Hulme 2009; O'Neill and Nicholson-Cole 2009). Sensing the scale of the challenge, the possible implications of unchecked climate change, and reacting to the challenge of connecting an abstract issue to a population largely unaware of the seriousness of the problem, fear and 'doom and gloom' messaging became a staple of the first wave of climate change campaigns in the 1980s and 1990s (Doyle 2011). Emotionally powerful (and strongly negative) graphics and images were frequently coupled with messages (correctly) emphasising the contribution that Western levels of material consumption were making to the problem. The combination of fear-inducing images, guilt-based rhetoric, and a generally negative outlook on the future characterised many of the early climate campaigns by NGOs such as Greenpeace (Doyle 2011).

These sorts of approaches have been challenged and critiqued from a number of perspectives. On the one hand, studies confirm the potential for fear to change attitudes or verbal expressions of concern, and under certain circumstances, actions and behaviours. But the impact of fear-based messages is context- and audience-specific. For example, those who do not yet realise the potentially 'scary' aspects of climate change need to first perceive themselves as vulnerable to the risks in some way in order to feel moved or affected (Das et al. 2003; Hoog et al. 2005). As people move towards contemplating action, fear appeals can help form a behavioural intent, providing an impetus or spark to 'move' from; however, such appeals must be coupled with constructive information and support to reduce the risk of feeling overwhelmed (Moser and Dilling 2007).

Much of what is known about how to strike the right balance in deploying fear-based messaging comes from public health campaigns, which are frequently faced with the challenge of conveying strongly negative risks into engagement that mobilises a behavioural response. Comparisons are often made with smoking, healthy eating, or reducing alcohol consumption, because (to a certain extent) scaring people into changing their health behaviours has 'worked'. As we explore in detail in Chap. 4, the interest in applying the principles of social marketing to public engagement with climate change is informed by the success this approach has had in the health domain (Corner and Randall 2011).

A key question, however, is the extent to which energy behaviours linked to climate change are 'like' health behaviours such as smoking. Drawing direct parallels between smoking, which is an isolated behaviour with direct and demonstrably negative impacts on an individual, their friends, and their family, and the distributed, diverse range of behaviours and social practices that collectively comprise an individual's carbon footprint is difficult. The lesson from this research is that unless carefully used in a message that contains constructive advice and a personal and direct link with the individual, fear is likely to trigger barriers to engagement with climate change, such as denial or dismissal of the problem. The danger is that fear can be disempowering – producing feelings of helplessness, remoteness, and lack of control (Stoll-Kleemann et al. 2001; Weber 2006; Moser and Dilling 2007; Lorenzoni et al. 2007).

This evidence is backed up by important lessons from campaigns on HIV/AIDs. Public communication of the threat of HIV/AIDs and associated behavioural changes were, at least initially, often framed in an extremely doom-laden way. Early UK campaigns in the 1980s were memorable for their fear-inducing images of cascading rocks giving way to shots of a tombstone or a looming iceberg. These public engagement methods were copied by governments around the world and are credited with significantly raising awareness of the threat from the virus. Analysis has shown that this fear-based approach was effective at attracting attention and was memorable, but was more persuasive for individuals who were *already engaging* in health-protective behaviours (Bourne 2010). For those engaging in risky behaviours, fear-based messages appear to have had unintended consequences including denial or a tendency to 'other' the issue.

Also directly relevant for climate change and energy messaging are analyses showing that apocalyptic messaging around risky HIV-related behaviours (that people have engaged with for a long period of time without harmful consequences) can lead to questioning of whether the message is accurate and the messenger is trustworthy (Witte and Allen 2000). To the extent that most people will have experience of engaging in 'risky' high-carbon behaviours without any obviously harmful consequences, an ongoing focus on fear and guilt in energy and climate change campaigns is clearly a problematic approach. But the nadir of this kind of approach did not arrive until 2010, in possibly the most dramatic public expression of a fear-and-guilt narrative in the history of climate campaigns.

The UK climate change organisation 10:10 brought together acclaimed screenwriter and producer Richard Curtis with national celebrities in the hope of producing an advert that would bring the issue to public attention while making people laugh.[2] However, while clearly intended as a parody, the video includes graphic scenes of school children's heads exploding (because they had failed to reduce their personal carbon footprints). It was met with such a backlash that it was pulled on the day of release, and for many, the message was lost and only served to confirm their negative stereotypes of environmentalists.

The lesson from public health campaigns and the academic literature for climate change engagement is not that campaigners should downplay the risks or avoid making people feel negative emotions. Fear-based messaging can be effective when it depicts a significant and relevant threat, as a short-term method of attracting attention and raising salience, for those already on the 'right path' to changing their behaviours, and when proportionate and constructive responses to the threat described are also identified. But it can also be counterproductive, and so this type of approach should be deployed with care in campaigns, learning the lessons that the literature on reducing risky health behaviours offers.

2.3   Mass Communication or Interpersonal Engagement?

The vast majority of public-facing campaigns on energy and climate change have been delivered through mass-media channels and advertising-based approaches (Corner and Randall 2011; Doyle 2011). As we discuss in detail in Chap. 4, there is huge potential in more participatory approaches to public engagement that are based on conversation and interaction, rather than 'messaging'. And learning from other contentious societal issues supports this conclusion.

As one of the pillars of modern healthcare, immunisation programmes require parents to take a preventative 'personal' action for both a personal and societal benefit (so in that sense are analogous to energy-reduction behaviours in response to climate change). The effectiveness of immunisation-based interventions has resulted in the virtual eradication of many life-threatening infectious diseases in populations across the world. Parents therefore rarely have personal experience of the threat posed by the diseases yet accept the necessity of vaccinations due to a combination of positive social norms and trusted medical advice. The importance of these factors in determining public understanding and accompanying action was keenly demonstrated by the anti-measles-mumps-rubella (MMR) vaccination controversy in the UK. In 1998, a now discredited piece of research indicated that there was a link between the vaccination and increased levels of autism among children.[3] The paper gained significant media attention, and vaccinations by 2003–2004 had fallen from 88 % in 1998 to 80 % nationally and as low as 61 % in some areas of London – well below the level needed for 'herd immunity'.

Learning from the controversy is instructive when reflecting on the way that climate campaigns have tended to be positioned and delivered. Pro-vaccination information was largely delivered using data and statistics, via government and medical officials (rather than through networks of parents and carers), and analyses demonstrated the importance of social networks in reinforcing parental understanding and propagating anti-vaccination beliefs (Petts and Niemeyer 2004). Parents were part of an evolving social interpretation of the risk which was often informed by a discourse of distrust of authority in general (Yaqub et al. 2014) and of pharmaceutical companies in particular (Weisbart 2015), rather than a lack of information. Combined with a lack of tangible salience around the diseases being inoculated against, a toxic cocktail for public engagement was created. Research reflecting on how lessons could be learned from the MMR situation recommended building trust among social networks, and developing a much clearer understanding of patient's needs and concerns, rather than delivering generic, data-driven, and top-down messaging – lessons that could equally be applied to energy and climate change campaigning (Petts and Niemeyer 2004; Downs et al. 2008; ECDC 2012).

Similar understanding has developed around anti-smoking campaigns. Although mass-media promotions are often the most cost-effective ways of 'reaching' large numbers of people, one-way communications have been most effective when combined with more interpersonal or community-based initiatives to support individuals in 'quitting' and to visibly shift social norms when positive behaviours are adopted (Mahoney 2010[4]). Other research, on the effectiveness of HIV campaigns, points to the importance of fostering environments where 'the voice of those most affected by the pandemic can be heard' (Panos London 2003), arguing that only when people become truly engaged in discussions and talking about HIV, does real individual and social change come about. Emerging approaches to public engagement with transgender issues in the US (dubbed 'deep canvassing') have produced remarkable shifts in attitudes, prompted by reflective and considered doorstep interactions (rather than billboard advertisements - Brookman and Kalla 2016).

As we argue in Chap. 4, the notion of widespread climate conversations is (in the context of past campaign strategies) a radical act. But the clear lesson from other 'difficult' or contentious social issues is that talking and participatory interaction is an essential tool in the box for deepening (as opposed to simply widening) public engagement. While many energy and climate change campaigns *involve* social interaction, face-to-face contact, and dialogue, this is rarely their main focus. More typically, this sort of approach is seen as a means to an end – to oppose a particular development or support a specific policy target. These sorts of targets are important and necessary, but they should follow (rather than precede) a process of public engagement.

2.4 POLITICAL POLARISATION AND THE 'BLAME GAME'

It is now well established that climate change is a politically polarised issue, at least in some key Western nations, where political orientation and ideology are among the most significant influences on public engagement (Hornsey et al. 2016). Those with right-of-centre political views are typically less concerned, more sceptical, and correspondingly less receptive to messages about climate change (Leiserowitz et al. 2015; Whitmarsh 2011). The usual explanation advanced for this relationship is that there is a conflict between conservative values – in particular around free market paradigms and individualism – and policies to tackle climate change (Corner et al. 2016).

But it wasn't always this way: ex-British Prime Minister Margaret Thatcher famously was an early advocate of the importance of climate change (Marshall 2015). The challenge of overcoming political polarisation on energy and climate change has been a major focus of our work at Climate Outreach, and we return in detail in subsequent chapters to this question. There has been a consistent failure to engage this large, global audience on climate change in a way that resonates with conservative values. While policy choices around energy use and energy technologies have frequently been framed in terms of values familiar to political conservatives – protecting energy security and resilience (to 'keep the lights on') or conserving the landscape (by scrapping onshore wind farms) – climate change has been noticeably absent from the conversation or, in some cases, actively marginalised and opposed. Our argument is that there is no inherent contradiction between centre-right values such as these, and engagement with climate change. But climate policies have not been framed in a language that connects with (rather than threatens) centre-right political perspectives. At least to some extent, then, the polarisation observed on climate change is a consequence of choices made about campaign strategies and the spokespeople (typically politicians, celebrities, and campaigners on the political left) who have defined the debate (Corner 2013a).

In recent years, as concern about continuing polarisation has grown, and campaigners have been forced to confront the fact that their 'standard' rhetoric and language is unlikely to deliver the broad-based support so urgently required for rapid societal progress on decarbonisation and climate change, interest has grown in finding more constructive ways of engaging across the political spectrum and reducing polarisation (Wolsko et al. 2016; Corner et al. 2016). Many campaigners now recognise that narratives about energy and climate change must be more inclusive if progress is to be made. But the temptation to demonise a convenient enemy remains and is identifiable in a relatively recent development in energy and climate change campaigning: divestment.

Taking its cue from previous campaigns to remove investments from racist or corrupt political regimes, it is easy to see why the fossil fuel divestment campaign has grown so quickly (Rowson 2015). Embraced by universities, religious institutions, and even entire city administrations, the concept has attracted a huge amount of support. The year 2015 saw the first 'Global Divestment Day'[5] celebrating the movement's successful move into the mainstream. However – as divestment advocates are the first to acknowledge – 'mainstream' is a relative term. It is estimated that more than

$50bn[6] has been removed from fossil fuel companies due to divestment campaigns. This is a striking success story, but it is also a fraction of what the industry turns over annually. The long-term power of divestment lies, therefore, in its potential to transform the social consensus on the merits of a fossil-fuelled economy and to create the political space for laws and legislation that will mean fossil fuels have to stay in the ground (Rowson 2015).

But – as is so often the case – climate change frustratingly doesn't easily fit the mould. Central to the rhetorical power of the divestment argument is an easily identifiable 'bad guy' (the fossil fuel industry) from whom the 'good folk' can dissociate. While it may be true that most of us don't personally quarry the earth for burnable carbon, almost everyone pays a quarterly energy bill straight into the coffers of the fossil fuel industry.[7] So whether we like it or not, most of us are complicit to some extent in the propagation of the system we are encouraged to divest from. To be clear, this doesn't make people hypocrites, but it is problematic for a simplistic portrayal of climate change as a battle between good and evil (because the enemy is literally within).

Painting fossil fuel support as immoral or even 'evil' is a strategy that could backfire when the values and perspectives of those outside of the divestment movement come into play. In our own research, we have found that people tend to react against an easy distinction between 'us' (fossil fuel opponents) and 'them' (the power companies). Because most people have no choice but to use and spend money on fossil fuels, there is a risk that the general public will feel more affiliation with 'them' than 'us'.

None of this means that the divestment movement does not have huge potential. But if divestment is to really go mainstream and start to uproot the foundations of the fossil fuel system, it is going to need wider support. The feeling of momentum that is currently providing buoyancy for the climate change movement must be shared by a larger group of the population. And for that to happen, there needs to be a much wider acceptance of the importance of climate change in the first place.

2.5 Maintaining Momentum: Movements Not 'Moments'

The international development sector struggles with many of the same communication issues that bedevil climate change campaigning: in both cases, the issues do not hold a great deal of salience to audiences in industrialised countries other than when a crisis hits. Public attitudes – at least in the UK – mirror perceptions of climate change, with a relatively high level of public awareness of the issue and general support for the need to 'act' (Henson et al. 2010). However, there is typically low understanding of the more subtle aspects of the causes and consequences of poverty, and a small (but vocal) group sceptical of the concept of providing support for international development from the domestic budget in the first place (BOND 2015). Public engagement is generally seen as one of the keys to success within the sector, and significant efforts have been focused on shifting public understanding. The most significant initiative in recent years was the Make Poverty History (MPH) campaign, which saw the biggest ever anti-poverty movement come together to increase awareness and pressure governments into taking actions towards relieving absolute poverty.[8]

MPH has been described as a 'masterclass' in public engagement.[9] Shifting the rhetoric away from traditional talk of targets, a range of more engaging and empowering messages were developed, with a palpable sense of momentum generated by celebrity backers focused on a key 'moment' (the 2005 meeting of the 'G8' nations). In a rare display of message discipline, member organisations merged their own organisational profiles and priorities. School children wore white bands, while churches and community centres were awash with the key messages of the campaign. The involvement of churches opened up a more bipartisan coalition as it had done in the US during the earlier Jubilee 2000 campaign (Sachs 2005: 343), allowing the religious right to view the issue in religious terms rather than traditional liberal terms (Church of England 2004). Close to a quarter of a million people took to the streets of Edinburgh for the flagship MPH march. Yet having seen a significant spike in public concern during 2005, all the gains soon dissipated, so that by 2009 levels had dropped to their lowest since 2003 (Darnton and Kirk 2011).

In analysing why such a significant campaign apparently had such limited long-term impact on public concern, there was general agreement that the slick communication approach resulted in 'wide but shallow' public engagement. As MPH co-founder Glen Tarmen stated, 'With hindsight, development NGOs recognise that we should have ensured that resources and strategies were in place so that the campaign went deep as well as broad' (McNeill et al. 2012). Kirsty McNeill, a member of the MPH coordination team, suggested that while broad, the coalition was largely made up of the 'usual suspects', commenting 'We need to get much better at forming unexpected strategic coalitions – not simply banding together with the nearest and most comfortable partners' (McNeill et al. 2012).

A similar analysis could be applied to the UK's 'Big Ask', one of the most high-profile public-facing climate change campaigns seen in any nation. A coalition led by Friends of the Earth generated significant public pressure to create a legally binding climate change law, and after numerous petitions, lobbies, music concerts, and demonstrations of public support, this was successfully achieved in 2008 (the UK Climate Change Act, 2008[10]). The scale of this achievement should not be underestimated, with the legislation setting a legally binding target of an 80 % cut to emissions by 2050. But for many campaigners, the creation of the Climate Change Act signalled 'job done' – with their focus shifting on to other pressing issues.

The Big Ask was highly successful at raising awareness for the time that it lasted. But arguably it did not deepen public understanding, instead focusing on policy-levers and emissions data, and made only limited efforts to build alliances with wider sectors. For campaigners and policy-makers caught up in the momentum of the Big Ask, there seems to have been an assumption that there would be a permanent shift in the political salience of climate change in the mid-2000s (Lockwood 2013), when in fact the wave of public interest reached in the run-up to the Copenhagen COP in 2009 came crashing down shortly afterwards, dashed partly on the rocks of a faltering economy (Scruggs and Benegal 2012). Campaigners demoralised by the underwhelming progress made at a summit widely billed as the 'last chance to save the world' quite literally stopped talking about climate change – and the public did too (Corner 2013b).

In the aftermath of the MPH campaign, there have been calls for deepening public engagement with development through a refocusing on the values that underpin the campaign rather than simply catchy messaging (Darnton and Kirk 2011). In response, some UK aid agencies have identified the need to adapt their approach to public engagement, to build a more positive set of narratives around development (based around the empowering frame of independence, as opposed to the traditional dependency frame). Throughout this book, we advocate for a comparable approach to public engagement with climate change – based on building engagement from the 'values-up' rather than from the 'numbers down'. But another key lesson from the MPH campaign for energy and climate change communicators is the need to maintain momentum, creating movements not 'moments'. Perhaps more than any other societal issue, climate change will not be 'solved' in the conventional sense of the word, and so communication and engagement is a continual and ongoing process. Packing up and going home after a successful campaign is not an option.

2.6   A Space for Radicalism

A central argument of this book is that public engagement is stuck in a rut precisely because energy and climate change campaigns have not built a wide and inclusive movement. Climate change remains socially and culturally associated with only a narrow band of activists, and this poses a barrier to making rapid and radical societal progress on decarbonisation. However, it is clearly not always possible or desirable for campaigners to be inclusive: there is a long and proud history of radical, direct action in the environmental movement and far beyond (Wall 2010). Indeed the most 'successful direct action in British history' took place in 1932, led by environmental campaigners seeking to open public access to the countryside, and such tactics have become a stalwart of environmental campaigns ever since (Norton 2014). By definition, these radical campaigns are not 'inclusive' or mainstream.

There is a large literature of analysis and commentary on radical activism, and it would stretch the remit of this book to review it in a detailed way. But the typical interpretation of how radical politics interacts with the mainstream is by opening up new political 'space', which the mainstream then ultimately inhabits. From the suffragettes to early (and at the time perceived as militant) gay-rights campaigners, there is no shortage of examples from recent and less-recent history that demonstrate how seemingly entrenched and immovable views and social norms can be quickly shifted (Tarrow 2011). Radical activism can be an important form of public engagement, mobilising communities of interest and creating compelling emotional narratives around bravery and commitment. Indeed, as public opinion tends to be conservative with regard to changes from the status quo, direct action is a key way of creating a new public discourse, bringing novel issues to public attention in dramatic ways (as the divestment campaign has attempted to for climate change). There is therefore a two-way relationship between those using radical actions and more mainstream public engagement (Doherty 2003; Schlembach 2011).

Several notable direct-action climate change movements have held protests and demonstrations – some more successful than others – over the past decade in the UK. Much of it has been focused on opposing specific high-carbon infrastructure such as airports, coal mines and 'fracking' sites (Nulman 2015). By directly targeting key sources of emissions, including power stations and airports, the activists created a very tangible sense of the scale of the transformations required in a way that policy discussions rarely do. Some of the occupations and demonstrations held at different locations around Britain have focused on demonstrating alternative ways of living, inviting others to share in a positive vision. Others have been more oriented towards disruption. And it is this latter category of climate activism where the question of how the activism was perceived by the wider world is most acute.

For example, the 'Plane Stupid' activists who disrupted Heathrow's schedules in July 2015 in a protest against the airport's proposed new runway were fighting an important battle.[11] Increasing the UK's high-carbon infrastructure is an irresponsible and short-sighted move if you care about climate change. But while they may have been clear in their own minds that high-end business flights were the target of their intervention, delays and cancellations to hard-earned summer holiday flights feed the negative stereotype of environmentalists who desire a return to a less comfortable, more austere pre-industrial age (Bashir et al. 2013).

Similarly, coal-fired power stations are an entirely valid target for activists seeking to speed up the process of decarbonisation. But while campaigners are quick to celebrate the closure of coal-fired power plants, they typically have less to say about the people who work there. Climate change policies and secure jobs are both non-negotiable. At one point this was reflected in climate change campaigns, with the (now dormant) Green New Deal arguing for a Bretton Woods-style reconfiguration of economic priorities and a 'carbon army' to provide jobs and achieve this aim.[12] But these kinds of arguments are no longer as visible – although they are still found among some parts of the trade union movement.[13]

Having a more human perspective on what decarbonisation means is crucial if radical activism is to complement rather than threaten mainstream concerns. As we argued earlier in this chapter, one of the tricky characteristics of the energy and climate change challenge is that there is no obvious 'enemy' against which to mobilise. Mining coal may be bad for the environment, but it is 'good' for local jobs (on a narrow interpretation of the word).

If radical activism really is to open up a space for the mainstream to move into, then radical activists must be seen as more than the 'usual suspects' advocating for a cause that is 'theirs' rather than something which wider society ought to engage with. If new audiences are going to be engaged, they will need to be able to relate to both the 'message' and the 'messengers'. If they cannot, the risk is that activism inflames social and cultural polarisation.

2.7   POSITIVE SIGNS

In this section, we briefly highlight two recent energy and climate campaigns that offer positive signals that campaigner strategies can move in a more inclusive direction, taking on board lessons from previous activism and learning from academic research.

The battle against a major new proposed oil pipeline – Keystone XL – in the US and Canada had been raging for years, before President Barack Obama relented in 2015 and scrapped the plans for this huge new piece of high-carbon infrastructure. The pipeline would have transported oil from the Canadian Tar Sands to the rest of North America, and its cancellation represents a significant victory. While many factors contributed to the success of the campaign (not least the tenacity and perseverance of the coordinators of the campaign, 350.org[14]), there was a striking difference between the Keystone campaign and many others like it: the names, faces, and social and cultural identities of the people opposing the pipeline were in many ways *not* the usual suspects. Ranchers from Nebraska and First Nations communities from Canada stood side by side, presenting an alternative representation of an 'environmentalist' to the wider public and contributing to the sense that the campaign was about 'us' not 'them'. A similar argument applies to a smaller, but still significant battle against a 'fracking' site in the north of England, led by self-styled Fracking Grandmas[15] (pensioners are not the first group of people associated with radical climate action). These different approaches are significant, because diversifying the identity of climate campaigners means expanding the social reality of climate change in the public mind.

There are also signs that the doom-laden and oppositional rhetoric of energy and climate campaigns is increasingly being abandoned. In one particularly clear example of this trend, the UK Stop Climate Chaos coalition responded to the post-Copenhagen blues by rebranding to the less apocalyptic The Climate Coalition and trialling a fundamentally different approach to their campaigning. Based on recommendations from research Climate Outreach conducted with four diverse groups of the UK public (including conservatives, members of trade unions, and community activists – Corner and Roberts 2014), the Coalition focused their messaging on one central concept: protecting the things that people love (or are passionate about) from the effects of climate change. The resulting campaign – For The Love Of[16] – is credited by the Coalition with significantly expanding their reach, allowing dozens of Coalition members (including WWF, Oxfam, RSPB, and the National Trust) to focus messaging for their members (and what they love) without losing the essence of the campaign.

There are certainly risks in taking a 'positive' approach too far. There is a danger that by focusing on the individual passions of relatively wealthy citizens in countries like the UK, the issue could be trivialised or the less palatable (but no less important) aspects of the problem sidelined.[17] And while the For the Love Of concept was (unusually) derived from research findings, any new approach should be carefully tested at scale to be sure of its usefulness and any unintended consequences. But a compelling, constructive, and emotive message based on connecting climate change to the aspects of people's lives that they care most about is a striking departure from pre-Copenhagen rhetoric about emissions cuts, unfamiliar notions of 'climate justice', and empty warnings of the 'last chance to save the world'.

2.8 Conclusion: Learn Lessons, But Be Prepared to Test Assumptions

There are dozens of examples of successful campaigns that have mobilised against a particular element of high-carbon infrastructure, and many that have achieved significant, game-changing policy 'asks'. But few have been able to spark and sustain a level of public interest and engagement – beyond the usual suspects – that is proportionate to the scale of the challenge.

Throughout this chapter, we have flagged the question of how comparable energy and climate change campaigns are to other societal issues. As we discussed in Chap. 1, there is a set of characteristics which make climate change – and developing a proportionate response to it – a uniquely difficult challenge. Campaigning on energy and climate change is tricky because climate change is an abstract and distant risk for most people, energy is embodied and embedded in our social practices and everyday lives, and decarbonising has implications for almost every facet of society. In many ways, then, climate change *is* 'different'.

Because climate change will not be 'solved' in the conventional sense of the word, there is no obvious point at which a climate campaign is finished and complete, and therefore very few glorious victories to be achieved. Major activist battles (for example, against the Keystone XL pipeline) may be won or lost, but qualitative shifts at a global scale are difficult to point to. Even when – as at the Paris UN summit in 2015 – a major break-through is achieved, the progress feels partial. Just like the problem they are designed to solve, energy and climate change campaigns and policies have an ephemeral quality, and it is difficult to gauge whether progress is genuinely being made.

But none of this changes the fact that there are better and worse ways for public engagement, communication, and campaigning to proceed and that practitioner strategies should be as aligned as possible with best practice from previous campaigns and the research base. And in this regard, the discussions in this chapter point to some illuminating conclusions.

Firstly, it is crucial to put a human face on energy and climate communication. Too many previous campaigns have led with technocratic targets, instead of human stories (Corner and van Eck 2014), and by following the precedent set around issues such as ozone depletion and acid rain, a model for policy and public engagement was established that seriously under-estimated the scale of the energy and climate change challenge. But who the 'human face' is really matters, and the tropes and (often negative) stereotypes of environmentalism have played a dominant, perhaps even stifling, role in framing the idea of climate change in the public mind. Although it is tempting to vilify a particular enemy (such as the fossil fuel industry) in climate campaigns, the notion of an easily identifiable 'bad guy' is questionable; when it comes to energy use, to some extent everyone is complicit.

Comparisons with other societal issues such as HIV/AIDs suggest that there is a place for using 'scare tactics' to engage the public with severe risks, but that this kind of approach only works under certain conditions. In particular, the risk must be personal and direct or activate resonant moral principles, with a constructive response available, or the fear evoked in the receiver of the message will quickly morph into dismissal or denial. In many ways, the analogy between climate change and particular health issues is not exact – not least because health challenges such as the uptake of vaccinations are concrete, measurable goals, where a perpetual process of public engagement does not seem necessary. But even for these more tangible targets, the evidence is clear: social networks and participatory dialogue are essential pieces of the puzzle.

Important lessons can be learned from campaigns – such as the Big Ask or MPH – that were judged as successful in terms of policy goals but later considered to have been a missed opportunity for sustained public engagement. Nurturing networks of engagement (rather than prioritising individualised, mass-market messages) is a theme we pick up on in Chap. 4, but the clear lesson from campaigns around immunisation and vaccination is that peer-to-peer communication holds a huge amount of sway and is likely to trump a dry, emotionless top-down strategy.

We have argued that while there is an important role for radical activism in promoting public engagement with energy and climate change, it must complement rather than threaten or exclude mainstream voices and interests. And we noted that there are encouraging signs in some recent initiatives of lessons being learned. The successful Keystone XL campaign was characterised by a notably diverse coalition of supporters, presenting a very different social and cultural reflection in the mirror of public opinion. And in the UK, The Climate Coalition has been pursuing a strategy that foregrounds the issues that people love and value and connects these to climate change. This is a radically different approach to bamboozling people with big numbers, scary statistics, and technocratic targets.

The central message of this chapter – and the first of the five principles advanced in this book – is that although we can learn lessons, we should also be careful not to assume an easy mapping between energy and climate campaigns, and other societal issues. In almost every way, climate change does not quite 'fit the mould'. This means that reflecting on assumptions and testing materials through applied audience research are crucial. Sometimes the intuitions of campaigners cohere with the recommendations of psychological and social research, but sometimes – as in the case of using fear-laden messages and imagery – they do not. So the most important first step for creating effective new approaches to public engagement is being prepared to test (and revise) assumptions about 'what works'. Social research takes time and resources, but it is a wise investment: campaigns that are not grounded in the growing psychological evidence base on climate change communication risk backfiring or (more likely) resonating only with the 'usual suspects'.

Notes

1. http://www.newleftproject.org/index.php/site/article_comments/a_ brief_history_of_climate_change. Accessed 23 June 2016.
2. https://www.youtube.com/watch?v=WjVW6roRs-w. Accessed 23 June 2016.
3. http://www.thelancet.com/journals/lancet/article/PIIS0140-6736 (9711096-0/abstract). Accessed 23 June 2016.
4. https://www.nice.org.uk/guidance/ph10/documents/expert-opinion-mass-media-interventions-summary2. Accessed 23 June 2016.
5. http://gofossilfree.org/wrap-up/. Accessed 23 June 2016.
6. http://gofossilfree.org/wrap-up/. Accessed 23 June 2016.
7. http://www.theguardian.com/sustainable-business/2015/mar/09/why-demonising-support-for-fossil-fuels-could-backfire-on-campaigners. Accessed 23 June 2016.
8. http://www.makepovertyhistory.org/takeaction/. Accessed 23 June 2016.
9. http://101fundraising.org/2015/03/make-poverty-history-lesson-what-can-it-teach-us-10-years-on/. Accessed 23 June 2016.
10. http://www.legislation.gov.uk/ukpga/2008/27/contents. Accessed 23 June 2016.
11. http://www.theguardian.com/uk-news/2015/jul/13/heathrow-disruption-climate-change-activists-claim-chained-runway. Accessed 23 June 2016.
12. http://www.greennewdealgroup.org/. Accessed 23 June 2016.
13. http://www.campaigncc.org/greenjobs. Accessed 23 June 2016.
14. http://thoughtfulcampaigner.org/learningkeystonexl/. Accessed 23 June 2016.
15. http://www.garstangcourier.co.uk/news/grandmas-pitch-camp-as-the-battle-against-fracking-looms-1-6788514. Accessed 23 June 2016.
16. http://fortheloveof.org.uk/. Accessed 23 June 2016.
17. https://newint.org/blog/guests/2015/12/01/darkening-the-white-heart-of-the-climate-movement/. Accessed 23 June 2016.

CHAPTER 3

The Building Blocks of Public Engagement

Abstract The climate change discourse continues to be dominated by big numbers and 'scary' statistics. But knowledge about climate change is not what drives public engagement. Values, worldviews, and political ideology are much more fundamental in shaping views about energy and climate change, and it is crucial to focus on these rather than policy goals or targets. Messages about energy and climate change should be framed around shared, communal values, not narrow economic self-interest. The second principle for public engagement is to *build public engagement from the 'values-up' rather than 'numbers-down'*.

Keywords Values · Worldviews · Political ideology · Science · Facts · Numbers

PRINCIPLE 2: PUBLIC ENGAGEMENT SHOULD START FROM THE 'VALUES-UP' NOT FROM THE 'NUMBERS-DOWN'

3.1   THE BIG NUMBERS OF THE ENERGY AND CLIMATE CHANGE DEBATE

At the heart of the challenge of communicating climate change is the fact that it is so frustratingly abstract. It must be captured, described, and communicated in order for us to interact with it at all. Our understanding of the climate is always filtered through an intermediary of some kind, whether this is a newspaper editorial, campaign materials, or simply the readings of a scientific instrument. Climate change always requires some degree of interpretation and translation, and so it is understandable that the original 'communicators' of climate change – climate scientists – have played a key role in shaping how people relate to the issue.

For obvious reasons, climate scientists have focused on measuring, monitoring, and quantifying the effect of human activities on the climatic system, producing a vast array of metrics with which to describe this relationship. These metrics have had a profound influence on the way that not only scientists, but also campaigners, policy-makers, and ulti-mately the general public relate to the issue. It is almost impossible to have a conversation about climate change without making reference to some of the 'big numbers' that define the debate.

To take just one example, one of the world's best-known climate change campaign groups (350.org[1]) is named after the level of atmo-spheric carbon dioxide that they (and many others) consider to be safe (specifically, a maximum of 350 molecules of CO_2 for every million molecules in the atmosphere in total). As we discussed in the previous chapter, many public-facing campaigns on energy and climate change have focused on percentage-based cuts in emissions of greenhouse gases (e.g. the 'Big Ask' campaign advocating for the UK Climate Change Act) or implored people to take 'action' against unfamiliar chemical equations (e.g. the UK government's 'Act on CO_2' campaign).[2] But are they a help or a hindrance for public engagement?

Perhaps the most famous (and arguably the most opaque) concept in the climate communication canon is the idea of a 2 degrees rise in global average temperatures (relative to the pre-industrial revolution average) as a 'dangerous' limit for climate change (Shaw 2015). After many years of intensive disputes, debates, and disappointments, the twenty-first meeting of the UN Framework Convention on Climate Change (UNFCCC) in Paris finally delivered a global agreement to keep temperature rises to 'well below' the 2 degrees target in 2015, with the aim of achieving 'net zero' emissions of carbon dioxide by the end of the century. It was the moment that the world had been waiting for, signalling a level of ambition that exceeded what many battle-weary campaigners believed was possible. While the hard work of translating the rhetoric into reality still lies ahead, on its own terms, Paris was a huge diplomatic success.

But behind the 'agreement' lies much that is still disputed, and it is these points of dispute that will determine how the lofty ambitions committed to paper in Paris will play out over the coming decades. Beyond the science-policy interface, most people don't think about the future by picturing a change in global average temperatures. It is crucial to ask how the radical carbon cuts necessary to achieve the aspirations of the Paris deal can be translated into people's lived experiences. What does 2 degrees mean?

In research conducted just before the Paris negotiations with members of the UK public, we found that people were baffled by the 2 degrees concept and puzzled that the challenge of climate change would be expressed in such a way (Shaw et al. 2015). In the only survey that has directly interrogated what people understand by the concept of 'dangerous' climate change (Carbon Brief 2013), a huge variation in responses was recorded. Only a minority perceived 2 degrees as indicating dangerous climate change – with most selecting 5 degrees or more as their best estimate (a level of global warming that would render large parts of the world completely uninhabitable). People understandably gauge temperature changes according to their everyday experiences, and a daily temperature fluctuation of 2 degrees is inconsequential, pleasant even – so why should they worry?

Our argument is not that people ought to have a better calibrated idea of what level of temperature rise is dangerous, but that average global temperatures are not a meaningful metric for public engagement. Despite the fact that the UN negotiations are the 'flagship' global events for climate change, very little thought has been given to how they are perceived or understood beyond the halls of the conference venues where the decisions are taken. Following the Paris agreement, the disagreements that remain are where the real decision making will now take place – disputes grounded in different values and played out in the familiar fight between differing worldviews and ideologies.

Despite the post-Paris rush to celebrate the end of the fossil fuel era by many climate campaigners, the truth may be a little less straightforward. In addition to the 'net zero' target, there are precisely zero mentions of fossil fuels in the final Paris text and zero indication of how the production of fossil fuels (as opposed to the emissions they cause) will be curtailed. Carbon neutrality could mean anything from abandoning fossils fuels altogether to relying on climate engineering technologies that are currently untested and likely to be unpopular among the public (Corner et al. 2012).

The notion that there is a single, dangerous amount of climate change has been criticised from many different perspectives (Shaw 2015), but technocratic targets like these would arguably not be so problematic if they were easier to peg to people's experiences. If the Paris agreement was grounded in a strong positive vision of what a sustainable future could be like – rather than an abstract target for average global temperatures – there would be less scope for the ambition of the deal to dissolve in the salty waters of national political pragmatism. But because the focus has continued to be on a somewhat arbitrary notion of 'dangerous' climate change, the 2 degrees limit is incredibly vulnerable to mission creep.

3.2   When Is a Consensus Not a Consensus?

The notion that there is a consensus among scientists about the cause of climate change has been central to the discourse for many years. More recently, academic analyses have attempted to put a precise figure on the level of agreement in the scientific community that human activities are responsible for climate change – with '97%' the current best estimate (Cook et al. 2013). Climate sceptics have challenged this figure, and there has been a mixed reception among scientists themselves to the suggestion that science could or should be reduced to a 'head count' in this way. But wherever the exact figure lies, it seems clear that there is a significant gap between the *actual* level of scientific consensus on climate change and public perceptions of the extent to which scientists are in agreement: in many countries around the world, members of the public dramatically underestimate the level of scientific consensus (Lewandowsky et al. 2013; van der Linden et al. 2013, 2014).

The reasons for the disparity between public and scientific opinion seem fairly clear. A central plank in the strategies of climate sceptic lobbyists has been to consistently undermine the idea that climate change is an established fact and to amplify and exaggerate the level of uncertainty about the relationship between human activity and changes in the climate (Oreskes and Conway 2012). 'False balance' in media reports has skewed how most people think about scientific agreement on climate change (Boykoff 2011). Scientists often focus on what they don't know before emphasising points of agreement – and these elevated levels of uncertainty are a barrier to engagement (Corner et al. 2015a).

And public perceptions of the extent of the consensus among climate scientists seem to really matter. As several recent studies have shown, if people can be persuaded that the consensus on climate change is greater than they had previously thought, then their levels of belief in climate change and concern about the issue increase. In fact, in a recent meta-analysis (Hornsey et al. 2016) of dozens of academic studies that have analysed the factors that predict belief in the reality and seriousness of climate change, judgements of the scientific consensus played a major role, leading some to dub acceptance of the scientific consensus as a 'gateway belief' on which other climate-related opinions are predicated (van der Linden et al. 2014).

But while these studies clearly show the value of communicating consensus information in a carefully controlled research setting, the findings are difficult to square with recent history. The authors of these studies recommend that reiterating the consensus is an effective (and even 'non-political') tool in the climate change communication box (Maibach et al. 2014). But scientists, campaigners, and politicians have relentlessly reiterated the fact that scientists agree that humans are changing the climate for the worse, and still the disparity between scientific and public opinion remains. In reality it is no more possible to pursue a non-political strategy of public engagement on climate change (especially in the US and Australia) than it is to issue a neutral statement about abortion or GM crops. In all cases, the science has political implications (Kahan 2012). The climate change consensus must always be communicated by someone, and that person will always come loaded with cultural baggage. Perception of the scientific consensus may well be a 'gateway belief', but the key to that gate is an understanding of the range of factors that determine how and why people respond to any statement about climate science: their values, political beliefs, and level of trust in different communicators.

This doesn't mean that it is impossible to communicate about the consensus effectively – simply that the scientific consensus alone cannot overcome deep-rooted divides that stem from differences in values, world-views, and political beliefs. In much the same way that commitment to the 2 degrees target for global average temperatures does not (in itself) distinguish between a world powered by 100 % renewable technologies, or one driven by nuclear power stations and the chemical removal of carbon dioxide from the atmosphere, the scientific consensus tells us little about the question that underpins public engagement with climate change: 'what does it mean for how we should live'?

Even if 100 % of climate scientists agreed that the world was ending next Tuesday, they could not tell us what we should do about it. This would remain a social and political decision, with the consensus offering only one piece of (very) relevant evidence. But the nebulous nature of the climate challenge is not easily reducible to neat percentages: the big numbers only get us so far.

3.3   MORE SCIENCE OR MORE POLITICS?

The temptation to shovel in 'more science' to deal with puzzling shortfalls in public empathy or engagement is nothing new. For many years, it was assumed that a lack of public interest in (or support for) a particular socio-scientific issue (e.g. GM crops) could be explained by a lack of knowledge about the issue in question (Hagendijk and Irwin 2006; Wynne and Irwin 1996; Renn et al. 1995). But the past decade has seen a concerted shift away from this so-called deficit model of public engagement. The deficit hypothesis has been discredited by empirical evidence – multiple studies have failed to find a straightforward link between a lack of knowledge and concern about an issue (Sturgis and Allum 2004). But it has also fallen out of favour for another reason; it embodies the old-fashioned idea that public engagement is a one-way process, rather than a dialogue (Rowe and Frewer 2005).

However, in climate change communication, a dogged commitment to 're-stating the evidence' persists – even though it is now well-established that knowing more about science does not seem to straightforwardly predict concern about climate change (Kahan et al. 2012). There is even evidence that higher levels of general scientific literacy can amplify ideology-based differences and further polarise those whose social and political views predisposed them to be sceptical, or accepting, of climate science (Kahan et al. 2012). Dan Kahan and his colleagues in the Cultural Cognition research group at Yale University found that differences in climate concern were sharpest for those with high levels of scientific literacy: scientifically literate sceptics were the least concerned about climate change risks, whereas science-savvy non-sceptics reported the highest levels of concern. The implications of this finding are important: scepticism about climate change is unlikely to be overcome by presenting 'more science', as sceptics who score the highest on science literacy tests are likely to be the most confident in their beliefs.

This message – that communicating well is not simply a question of finding the perfect graph or the killer statistic – is a critical one for understanding how to develop a more effective approach to public engagement with energy and climate change. The facts and figures of climate science slot into a complex social, political, and moral context that exerts a formidable influence on whether, and to what extent, people let the risks of climate change sink in. The lesson of this body of research is not that the facts are unnecessary, simply that they are insufficient for public engagement. Being right is not the same as being persuasive, and the 'big numbers' of the climate change and energy debate do not speak to the lived experience of ordinary people going about their daily lives; they tell us nothing about the social, economic, and political dynamics that have created the world in which we live today and how we can adapt it to a changing climate. A societal shift on the scale required will only come from the values-up, not from the numbers-down.

3.4   COMMUNAL VALUES: THE BUILDING BLOCKS FOR CLIMATE CHANGE AND ENERGY CAMPAIGNS

Public engagement with energy and climate change is not reducible to a simple rule of thumb, but some aspects of human psychology are more fundamental – and explain more variation in attitudes and behaviours – than others (Hornsey et al. 2016). People's attitudes on different topics may morph and shift over time; they may switch allegiances between different political parties and candidates, their aesthetic preferences may wax and wane as they mature; individuals may even express quite distinct 'selves' in different social situations and with different groups of people. But there are certain aspects of people's psychological make-up that are relatively consistent and which form the core of their identities – and central among these are values.

There are multiple definitions of what a 'value' is, but the typical definition of a 'guiding principle in the life of a person' provides a good starting point (Schwartz 1992). Values are distinct from beliefs or 'attitudes', in that they are relatively stable and are brought to bear on many different situations, whereas attitudes are more malleable and may be situation-specific. While public surveys tend to capture and compare public attitudes towards energy and climate change (or other issues), values are the 'bedrock' on which specific attitudes are founded (Maio 2015).

Through a programme of research that has spanned several decades, 44 nations, and over 25,000 respondents (Schwartz et al. 2012), there is now a very robust body of evidence on values. The work of Shalom Schwartz and his colleagues in particular has identified 56 'universal' values that can be divided into distinct clusters which vary along two dimensions. These two dimensions are 'openness to change' versus a desire to conserve/respect tradition; and 'self-transcendence' (i.e. values which go beyond self-interest, such as altruism or forgiveness) versus 'self-enhancement' (i.e. self-focused values such as power, ambition, and materialism). Although people possess a range of different and sometimes conflicting values, it is tricky to make a decision based on values that are on opposing ends of a dimension. The Schwartz values model is visualised in Fig. 3.1:

This basic concept has been taken in slightly different directions by other researchers. Some have identified three broad clusters of values – egoistic (i.e. self-focused), biospheric (i.e. environmentally focused), and altruistic (i.e. others-focused; De Groot and Steg 2008) – or described a set of 'postmaterialist' values that have emerged in industrialised economies which have experienced growing affluence and economic/political security (Inglehart 2008). Others have focused on the notion of 'moral foundations' rather than values, arguing that ideas about basic moral principles such as avoiding harm, fairness, and loyalty are the best way of understanding differences between people and predicting how they will think and feel about different societal issues (Haidt 2007; Markowitz and Shariff 2012).

Some research has introduced a distinction between 'intrinsic' (e.g. for a sense of well-being) and 'extrinsic' (e.g. for a monetary reward) values, as motivations for behaviour in different contexts (Sheldon and

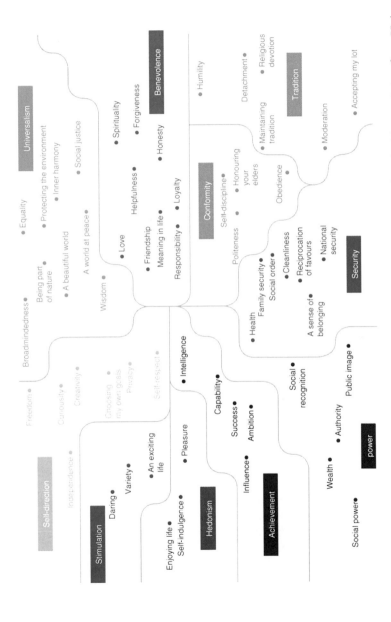

Fig. 3.1 A visual map of the Schwartz circumplex model of human values (Reproduced with permission from Holmes et al. 2011; based on Schwartz 1992)

Nichols 2009). Another strand of thought comes from anthropology (Douglas and Wildavsky 1982) and according to this approach, values (more commonly called cultural worldviews) exist on two cross-cutting dimensions and describe people's favourability towards different societal arrangements. The first dimension, 'hierarchy-egalitarianism', refers to people's preferences around how equitably resources are distributed, and social power relations. The second, 'individualism-communitarianism', relates to the question of whether individual interests should be subordinated to collective ones.

We describe these different approaches here not because we want to labour the differences between them, but because the basic idea that there are distinct, measurable values that people across a range of cultures endorse to a greater or lesser extent, and which predict people's opinions on a range of topics, has withstood a lot of scrutiny. As we explore in detail in the next chapter, the way that different values are used when communicating about energy and climate change matters, because promoting or 'priming' one type of value (e.g. by talking about the economic rationale for energy saving – a self-enhancing value) is likely to weaken the prominence of opposing values (e.g. the environmental benefits of energy saving – Crompton 2010). And on the question of how values shape views about climate change, there is a great deal of convergence between different academic theories.

People who favour self-transcending values are more likely to be concerned about climate change and support climate policies (Corner et al. 2014). People who hold communitarian views, and prefer egalitarian forms of social order, are less likely to be sceptical about climate risks (Kahan 2012; Zia and Todd 2010). People who identify with 'intrinsic' rather than 'extrinsic' motivations are more likely to care about climate change (Kasser and Crompton 2009). The research is clear: certain types of values are consistently associated with positive engagement with climate change, while others are not (Corner et al. 2014). For simplicity, we refer to these as *communal values* from this point onwards, and (as we explore in the next chapter) there is a clear choice for campaigners to make in terms of the types of values they build their campaigns on.

Self-enhancing values, individualistic worldviews, and extrinsically motivated behaviour are more commonly associated with the Right than the Left of the political spectrum, and so there is a well-established relationship between political ideology and engagement with climate change (Sheldon and Nichols 2009; Kahan 2012). Political conservatism predicts scepticism about climate change, both among individuals (McCright and Dunlap 2011) and in the media (Painter 2011), particularly but not exclusively in English-speaking countries. In the UK there is a direct relationship between voting for the Conservative Party and scepticism about climate change (Whitmarsh 2011), with an even more polarised partisan pattern in the US and Australia (Hornsey et al. 2016). The usual explanation advanced for this relationship is that there is a conflict between conservative values – in particular around free market paradigms and individualism – and policies to tackle climate change.

We return to the challenge of overcoming political polarisation on energy and climate change in the next chapter, where we argue that there is no inherent contradiction between *some* of the values that define conservative belief systems and engagement with climate change (at least, on the centre-right of the political spectrum). Communal values are not exclusive to one side of the political spectrum. But there has been a consistent failure to engage this large, global audience in a way that resonates with the elements of political conservatism that are *not* in conflict with a proportionate societal response to climate change. Climate policies have not been framed using language that connects with (rather than threatens) conservative political perspectives.

Our discussion of values, worldviews, and political ideology is designed to illustrate that they are fundamental building blocks for public engagement with climate change and that any strategy which seeks to widen and deepen public engagement must begin with these basic concepts. This does not mean that there are not a plethora of other influences – individual, social, and structural – that are constantly shaping public opinion, some of which we explore in subsequent chapters. People don't (or can't) always act in line with their values – the infamous 'value action gap' – but this does not undermine their importance. Over a broad enough range of situations, values are pretty good predictors of attitudes and attitudes are still pretty good predictors of behaviour (Maio 2011). And the communal values that underpin positive engagement with climate change are actually widespread – perhaps more so than is commonly realised.

One long-running source of evidence on the values of European citizens is the European Social Survey (ESS). This survey is carried out every two years with a large, representative sample of the European public (typically around 2000 people in over 30 countries). Among many other things, it includes questions on the values people hold. A report by the Public Interest Research Centre (PIRC) on European values, and how they relate to attitudes and behaviours linked to human rights and equality, analysed the ESS data (Blackmore et al. 2014). The top two most popular values across Europe as a whole were both communal ones – 'benevolence' (kindness) and 'universalism' (a recognition of the rights and welfare of all people).

The ESS findings are corroborated by other data. In a recent survey of the UK population, Tom Crompton and his colleagues compared the importance that people placed on different types of values to their sense of the values that other people in the UK held (Common Cause 2016). There was a striking difference in the values that people identified with themselves (with a clear majority favouring communal options) and the values they attributed to others (overwhelmingly self-focused). So while there are widespread misconceptions about the prevalence of the values that underpin positive engagement with climate change, in fact they are relatively common and popular. Our argument in the next chapter is that it is possible to build bridges between the wide range of communal values that diverse members of the public hold and a proportionate societal response to climate change – across the political spectrum.

There is also a significant amount of overlap between the values that people hold and the values they wish to see reflected in the energy system. In a comprehensive analysis by researchers at Cardiff University of public perceptions of the entire energy system in the UK – from low-carbon technologies to demand-side energy management – a core set of values that people associated with positive energy-system change was identified (Parkhill et al. 2013). These values included the protection of nature, fairness, and respect for the autonomy of individuals, a positive contribution to future well-being, efficiency, affordability, the avoidance of waste, and long-term thinking, as Table 3.1 shows.

The important insight from this study was that these were principles that people used to evaluate a whole range of different energy-policy options, as well as the desirability and feasibility of individual lifestyle changes. Just as people's personal values predict their views on a whole

Table 3.1   Core public values that underpin views about the energy system in the UK

Principle/Value		Description
Reduced energy *use* overall Reduced use of *finite* resources		Reducing overall energy usage while simultaneously reducing the use of finite resources (as compared to the current state) will have positive consequences in terms of attaining the values outlined below
Efficient and not wasteful	Avoiding waste Efficiency Capturing opportunities	A system that does not involve wasting and/or produces waste products and that is efficient. A system that does not waste opportunities arising from energy system change, and capitalises on the resources and capacities of the UK
Environment and nature	Environmental protection Nature and naturalness	A system that uses and produces energy in an environmentally conscious way and does not unnecessarily interfere with or harm nature
Secure and stable	Availability and affordability Reliability Safety	A system that ensures access to energy services both in terms of availability and affordability. A system that is reliable and safe both in the production and delivery of energy services
Autonomy and power	Autonomy and freedom Choice and control	A system that is developed in ways that do not overly threaten autonomy, infringe upon freedoms, or significantly compromise abilities to control personal aspects of life
Just and fair	Social justice Fairness, honesty, and transparency	A system that is developed in ways which are mindful of implications for people's abilities to live healthy lives. A system that is fair and inclusive and where all actors are honest and transparent about their actions
Process and change	Long-term trajectories Interconnected Improvement and quality	A system that is developed with a focus on the long-term trajectories being created, that takes into account system interconnections and interdependencies, and that represents improvement both in terms of socio-technological advances and quality of life

Reproduced with permission from Parkhill et al. 2013

Notes: In the table, the column on the left (Principle/Value) lists the principles and values that make up the value system. Each value or principle is accompanied by a brief description (right column). Naturally all of the values and principles are linked rather than mutually exclusive. As such they are grouped together according to connected meanings. Each set of values is then discussed in more detail alongside the table. It is in this narrative that we explore how these values are interconnected and what aspects of energy system change they relate to (and how) range of different social issues, so the values that people want to see reflected in the energy system drive their opinions on seemingly disparate energy policies. And these values are overwhelmingly communal in their focus.

3.5   CONCLUSION: PUBLIC ENGAGEMENT ON CLIMATE CHANGE SHOULD START FROM THE VALUES-UP, NOT THE NUMBERS-DOWN

The physical sciences have played an essential role in bringing climate change to the world's attention and provided a vital piece of the puzzle for ensuring our response to it reflects the objective realities of climate change. Temperatures really have risen, and the oceans really are more acidic, whether we want this to be true or not. No credible attempt to communicate climate change should perpetrate falsehoods or ignore the facts. And it is crucial not to lose sight of what the scientific data and 'big numbers' reveal: accelerating climatic changes that, unchecked, will have profoundly negative consequences for every aspect of human society. But because of the central role of scientific analyses and tropes, climate change communication has been dominated by a focus on 'big numbers' rather than a vision of the future, and changes in the energy system are usually quantified instead of 'imagined'.

In the aftermath of the UN agreement in Paris, with the biggest of the big numbers (2 degrees) now finally enshrined in international policy, there is a pressing need to step up our commitment to public engagement and translate the techno-babble of international policy negotiations into a language that ordinary people can connect with. Finding common ground on these more contentious topics is where the energies of climate communicators are best placed now that the skeleton of a sustainable world has been assembled.

The facts of climate science are like a dictionary: they provide the basic vocabulary. The real challenge is in weaving poetry and prose to inspire people to care about the problem. Fundamentally, this means engaging with people's values. Until the energy and climate debate resonates at this level, pointing to a row of nodding scientists and expecting this to catalyse public concern is not going to get us far – no matter what the 'magic number' attached to the consensus is. The abstract targets of international policy are great for negotiators but are not a meaningful currency for communication. Public engagement should be approached from the 'values-up' rather than the 'numbers-down'.

In the next chapter, we review the evidence on how to achieve this goal – the language and frames that can help to build narratives about climate change that resonates with the hopes and aspirations of more than just a narrow band of committed climate advocates.

Notes

1. http://350.org/. Accessed 23 June 2016.
2. http://webarchive.nationalarchives.gov.uk/20101007164856/; http://actonco2.direct.gov.uk/home/about-us.html. Accessed 23 June 2016.

CHAPTER 4

Language, Frames, and Narratives

Abstract Most communication happens via anecdotes and stories, not graphs and statistics. Although there are limitations to the impact of 'one-way' message-based communication strategies, identifying the right linguistic tools and narratives for starting more productive climate conversations is a critical aspect of effective participatory engagement. There are no 'magic words', but there are better and worse ways of starting climate conversations. Framing messages to engage with diverse communal values is important, and narratives about climate change that can engage beyond the 'usual suspects' and across the political spectrum can shift climate change from a scientific to a social reality. The third principle for public engagement is to *tell new stories to shift climate change from a scientific to a social reality.*

Keywords Words · Language · Frames · Narratives · Stories · Social reality

© The Author(s) 2017
A. Corner, J. Clarke, *Talking Climate*,
DOI 10.1007/978-3-319-46744-3_4

PRINCIPLE 3: TELL NEW STORIES TO SHIFT CLIMATE CHANGE FROM A SCIENTIFIC TO A SOCIAL REALITY

4.1   WARM WORDS?

The trappings of scientific and technocratic communication – graphs, charts, statistics, and projections – have played a central role in the energy and climate change debate thus far. But the majority of everyday communication takes place via stories, anecdotes, metaphors, colloquialisms, and the sharing of images, and these stories are never entirely 'neutral'. Intentionally or not, all information is 'framed' by the context in which it appears. The same information, when given a different label or title, presented by a different messenger or when linked to a particular theme or idea, can be perceived very differently (Lakoff 1990). Language matters, and a considerable research effort has been dedicated to documenting the effects of different linguistic choices on public engagement with energy and climate change (Nisbett 2009).

At the most basic level, there is an ongoing debate around which of the terms 'global warming' or 'climate change' is a more effective communicative tool. Some US research has found that 'global warming' creates a stronger sense of threat, proximity, and a desire for action than 'climate change' (Leiserowitz et al. 2014; Topos Partnership 2009). British research has also found the term 'global warming' to be more emotionally engaging (although the idea of a 'warmer' country is intuitively appealing for many British citizens – Whitmarsh 2009). However, one US study found that Republicans were more likely to endorse the reality of 'climate change' than 'global warming' (Schuldt et al. 2011).

Periodically, there have been attempts to introduce new phrases, including 'our deteriorating atmosphere' (Western Strategies and Lake Research Partners 2009), 'global weirding', and 'global climate disruption' (Revkin 2008). In recent years, the concept of 'carbon pollution' has begun to dominate the US President Barack Obama's climate change and energy speeches – a deliberate attempt to overcome the problem that both 'global warming' and 'climate change' are frustratingly abstract and unsituated terms. As we discuss below, there has been a general recognition of the need to frame carbon emissions with regards to their (more tangible) health implications (e.g. Maibach et al. 2010), and the growth in popularity of 'carbon pollution' in the US seems to be a direct result of this.[1]

In fact, the aim of making climate change more 'tangible' has been the focus of a range of studies aimed at reducing – or at least managing – the so-called psychological distance of climate change (Isaksson and Corner 2016). Most people (certainly in Western countries) tend to think of climate change as something that happens to other people and in the future. There is a consistently documented tendency for respondents in surveys to assess the risks of climate change in inverse proportion to their proximity to themselves and their families: most people do not feel personally at risk from climate change, and (thankfully) most people do not directly 'experience' or 'encounter' climate change in their daily lives. Moreover, the inherent uncertainty in climate projections allows for wishful thinking (Spence et al. 2012). People are more likely to be willing to act on climate change if they think that it will impact them (or people they care about and who are similar to them) in the near future (Isaksson and Corner 2016). So language that positions climate change as something worthy of an individual's personal concern, or as something that is in the interests of a community to engage with for their own sake, has become a major focus of research.

In general, 'localising' climate change (for example by communicating the current and future impacts of sea-level rise on local communities) has been shown to help reduce the sense that climate change is a 'distant' issue for particular audiences (CRED & ecoAmerica 2014; Spence et al. 2012). But localising strategies are not a panacea for building public engagement, and several studies have suggested that there are risks in 'over-localising' climate messages too (McDonald et al. 2015). In a nutshell, while messages that talk about the global and abstract dimensions of climate change may leave people feeling that 'climate change isn't relevant for me', messages about climate change which focus only on hyper-local reasons for caring about the issue may inadvertently trigger the equally unhelpful response – 'climate change only matters to the extent that it is impacting me' (Brugger et al. 2015). Plus, research on using 'local' imagery as a tool for engagement has also found that it can be disengaging, because 'it will only affect locals and is not as much of a global issue' (O'Neill and Hulme 2009). There is a danger in framing approaches which trivialise the interaction between people and climate change – in images or in written materials.

Another key focus of research on framing messages about climate change is the perpetual challenge of communicating uncertainty (Corner et al. 2015a). Although the basics of climate change science are now well established and no longer in dispute, there are innumerable issues (for example, how regional impacts of climate change will manifest in decades to come) where much uncertainty remains. Just like in any area of complex science, uncertainty is a feature of climate change that will never go away: it is not an enemy of climate science that must be conquered – it is a stimulus that drives research forward. The fact that we have imperfect knowledge about climate change should only increase our motivation for taking preventative action against uncertain risks. But unlike in economic forecasts (which are widely accepted despite sometimes proving wildly inaccurate), or medical diagnoses (which everyone accepts contain an element of chance), uncertainty has become an argument for discrediting and doubting climate science and for delaying policy responses.

Partly, this is because political lobbyists opposed to societal action on climate change (so-called Merchants of Doubt – Oreskes and Conway 2012) have intentionally manufactured distrust around the science of climate change, exaggerating areas of uncertainty while downplaying areas of strong consensus and agreement. But even without such distorting influences, the communication of uncertainty is still a formidable challenge. One difficulty is that science is often represented by the media as a series of definite facts and figures: either 'unprotected exposure to UV rays causes skin cancer' or it doesn't. But in reality, scientists work with probabilities (so the truth is that unprotected exposure to UV rays makes skin cancer more *likely*). Similarly, in schools, science is taught as a series of 'answers' rather than as a method for asking questions about the world. And as a consequence, people seem to have different expectations about uncertainty in science, relative to 'everyday' situations where uncertainty is more likely to be taken as a given. Because of these challenges in communicating uncertain information, there has been a lot of interest in how uncertain statements about climate change can be 'reframed' to be more engaging for a non-scientific audience.

For example, some research has recommended reframing uncertainty information using the closely related concept of 'risk', which as the language of the insurance, health, and national security sectors, most people are more familiar (Pidgeon and Fischhoff 2011; Painter 2015). Everyday examples of risk management offer useful comparisons and analogies (e.g. the idea that we all take out house insurance against fire damage, even though the chance of this happening is miniscule), which provide a firmer linguistic foothold than focusing on uncertainty (Corner et al. 2015a).

However, while the substitution of more understandable words and terms, and useful analogies to other areas of life where we seem more able to grasp the notion of risk-based decision making are helpful, there is only so much that this kind of 'tweaking' can be expected to do (Shaw et al. 2016). Many key phrases and terms have now been in the public domain for a long time and therefore already have strongly embedded social and political meanings for many people (Marshall 2014d; Villar and Krosnick 2011). And despite some concerted attempts at introducing a more 'risk-based' register into the reporting around the release of key scientific reports (Painter 2015; Painter and Gavin 2015), there is not a great deal of evidence that media reporting has adopted a different strategy.

A similar problem exists for visual communication: in the same way that some linguistic terms have become inextricably embedded in the public discourse, so certain visual frames have come to be firmly associated with climate change in the public mind. In research with members of the public in the UK, Germany, and the US, we found that despite widespread cynicism about the value of 'classic' climate images such as polar bears on melting ice, or smokestacks, these images were readily and rapidly identified as visual shorthand for climate change (Corner et al. 2016). After more than twenty years of dominance, the longstanding positioning of climate change as a predominantly 'environmental' (rather than human) issue has largely resisted attempts at reframing it.

If campaigners and communicators want to tell a more compelling visual or verbal story about climate change, some very deeply ingrained associations will have to be overcome. And there is a growing body of work that has attempted to do exactly this – broadening the social reality of climate change by highlighting the diverse areas of people's lived experiences that it relates to, in particular around the health consequences of climate risks (and health benefits of addressing them).

4.2   What If It's a Big Hoax and We Create a Better World for No Reason?

Many of the negative societal impacts of climate change could just as easily be described as 'health risks': flooding and droughts, unpredictable and extreme weather, and the spread of airborne diseases as temperatures fluctuate are just a few examples (Luber and Lemery 2015). The potential advantages of using a 'public health' frame to talk about climate change have been stressed by a number of authors in recent years. As something that everyone has an obvious and immediate stake in, public health is an important 'interpretive resource' for people when engaging with climate change (Nisbet 2009). Communication about climate change that connects to health problems which are already familiar and seen as important (such as heatstroke and asthma) can make the issue seem more personally relevant (Nisbet 2009). Ed Maibach and his colleagues (Maibach et al. 2010) found that portraying climate change in ways that affirmed the health benefits of taking action on climate change made the issue seem more personally significant and relevant to participants in a US study. Another US study (Myers et al. 2012) tested differently framed messages about climate change, and the public health frame elicited the most positive responses. A public health frame thus offers the prospect of shifting the terms of the debate from 'concern about climate change' to 'concern about public health', which is likely to be a priority for audiences irrespective of ideology and political outlook (Maibach et al. 2010).

Because of the broad overlap between pro-environmental and pro-health measures, many programmes primarily aimed at reducing energy consumption (e.g. through better home insulation) or promoting renewable technologies place a central emphasis on the health 'co-benefits' of these measures.[2] The language of 'dirty' fossil fuels and carbon pollution, positioned in direct opposition to clean, fresh, and healthy renewable technologies, has become increasingly popular among communicators. And this approach of emphasising the health gains of low-carbon policies is consistent with other work showing that a 'gain' frame (emphasising the benefits of action, rather than the negative consequences of not acting) produces more positive attitudes towards tackling climate change (Rabinovich et al. 2010; Spence and Pidgeon 2010). Other studies talk about the importance of 'motivational' frames (that highlight the collective benefits of climate mitigation) rather than 'sacrificial' ones for building public engagement. In one Canadian study (Gifford and Comeau 2011), exposure to motivational messaging was found to increase people's perceived 'self-efficacy' to adopt pro-environmental behaviours – that is, their sense that they can personally make a difference.

Even among an audience of so-called climate change 'deniers', messages that identify positive outcomes of mitigation efforts (such as improvements in social welfare and creating a society where people are more considerate and caring, or that focus on the economic and technological development that climate policies can bring) seem to be more effective (Bain et al. 2012). One study with the US conservatives found that people were more favourable towards environmental messages when these focused on pollution and the 'purity' of the natural environment – rather than the more conventional set of arguments about a moral responsibility to avoid harm (Feinberg and Willer 2013).

Given that policies to combat climate change and lower energy use really do offer a wide range of unintended benefits, the evidence from the psychological research seems clear: while there is nothing to be gained by downplaying the seriousness or urgency of climate change, talking about the many ways in which climate policies will benefit society makes a lot of sense.

There is, of course, a balance to be struck between identifying and promoting the benefits of climate policies, and the practice sometimes known as 'brightsiding' (i.e. putting a positive spin on any situation, no matter how negative it really is). Partly in response to the prevalence of doom-laden and apocalyptic messaging around climate change, some communications specialists began advocating 'selling the low-carbon dream' rather than lamenting the nightmare of a climate-changed future.[3] However, the term tends to be used pejoratively by those who – in our view correctly – see a relentlessly (and inauthentically) positive framing of climate change as no more helpful than an overly pessimistic assessment of the challenge (Spratt 2012). Recent research (Fielding and Hornsey 2016) has also suggested that 'good news' may not always be as motivating as some communicators assume. In a study with US, Australian, and UK participants, two messages about the recent 'slowdown' in global emissions of carbon dioxide were tested – one with a positive and one with a negative slant. They found that relative to the more pessimistic message, the optimistic message reduced participants' sense that climate change represented a risk to them. Rather than increasing their sense of self-efficacy, it seemed to breed complacency. So just as with fear-based messages, the key lesson seems to be providing people with a constructive 'action' or response they can take, no matter whether they are responding to good or bad news on climate change. An authentically positive framing of climate change is one that is constructive, not one that claims 'everything will be fine'.[4]

4.3   MAGIC WORDS

In the same way that starting with the 'big numbers' of climate change is not as effective as grounding communication strategies in the values that people hold, there is a limit to tinkering with and exchanging single words and phrases, in order to emphasise the co-benefits of climate policies. The studies reviewed above produced only modest effects on people's concern about climate change or support for dealing with the problem. Substituting different terms in communications that ultimately convey a similar underlying message is likely to have a limited impact in the messy, noisy world beyond the laboratories in which most research on message framing takes place. If people are encouraged to consider climate change as a secondary concern to other more important issues like their health, or a sense of security (rather than inseparable from them), then the same relatively shallow level of engagement remains: there may be co-benefits to engaging with climate change, but climate change itself is not something worthy of their deeper attention.

Because of this, conceptualising climate change as something that offers co-benefits is likely to be vulnerable to attempts at dismissing or undermining these benefits, by opposing voices. Plus, in our own work at Climate Outreach, we have found that people tend to be uncomfortable with (or even hostile towards) policy jargon such as co-benefits or 'win-win solutions' (Marshall et al. 2016). Real progress would be represented by climate change simply being *equated* with the various positive impacts on health and security and well-being that a lower-carbon world will bring; externalising the co-benefits of climate policies creates an unhelpful distinction between climate change and people's lived experiences.

A recent comparison of the effectiveness of several different framings of climate change messages – relatively rich information presented to American participants in an online study as realistic-looking newspaper articles – makes this point emphatically (McCright et al. 2015). People in the study viewed a newspaper article that talked about climate change as either an economic opportunity, a matter of national security, a question of Christian 'stewardship', or a public health threat. In common with other studies reviewed above, only a weak positive effect of the framing techniques was observed on people's views, with the 'economic opportunity' and 'national security' frames producing somewhat more positive views about the importance of ambitious US mitigation policies. But unlike most framing studies, the researchers also included a 'counter-frame' that encompassed anti-climate change or 'denial' themes. The presence of this counter-frame consistently undermined the impact of the positive frames, suggesting that even through framing-based approaches can produce measurable shifts in public views, they may be fragile or temporary. Precisely because public engagement is relatively shallow, it is vulnerable to the 'misinformation' campaigns of those groups and individuals who are opposed to societal action on climate change.

Another recent paper (Bernauer and McGrath 2016) reported on a series of experiments that also poured cold water on the importance of reframing climate messages. In a study where people were asked to read messages about climate change framed using different concepts (including health), they found no clear evidence that any of the frames had much effect. In a despairing opinion piece[5] responding to the new study, leading environmental journalist David Roberts argued that spending time (and funders' money) on reframing climate change messages was a dead end. Roberts dismissed the value of message framing, suggesting that magic words would not alter people's longstanding beliefs and perspectives, which are grounded in deep-rooted (and therefore unchangeable) values and worldviews.

Roberts was right to caution against a simplistic understanding of message framing. There are no magic words, and it is naive to think that people's attitudes can be transformed through exposure to a few 'reframed' messages about climate change. But he was mistaken to argue against the wholesale abandonment of the value of message framing in two important ways.

Firstly, there is no inconsistency between 'message framing' and people's deep-rooted values and worldviews. As we show in the next section, messages can be framed to speak to these values – to reflect or even reinforce existing beliefs, rather than trying to change them. The challenge is not to overturn people's values, but to diversify the social and cultural meaning of energy and climate change, so that the issue is 'owned' by people with a diverse range of values. If the right 'frame' is used by a trusted communicator – someone who shares the values of audience – some of the building blocks of effective communication are in place. There is a big difference between message framing presented within a neutral academic context and language which is delivered by a trusted peer or opinion former. The academic research does not show that frames do not work, but simply that they are insufficient in the absence of other social cues for people to take the message seriously.

Secondly, the value of reframing climate change comes not through 'one shot' advertising messages, tweets, or behavioural nudges, but through using the most effective language as a tool for starting a dialogue.

Using the right language is about starting a productive dialogue, not 'winning an argument'. There may be no magic words, but there are definitely better and worse ways of starting a conversation. Framing matters, because starting a conversation with someone on terms they are comfortable with is the first step to building – and sustaining – their engagement. Message framing (as part of a process of participatory public engagement) is not a waste of time, even if a handful of reframed messages cannot (on their own) overturn the temptingly reassuring arguments of climate sceptics.

It is noteworthy, though, that while sceptical campaigns often involve outright falsehoods, they also speak effectively to people's values and core beliefs and tell a compelling story that resonates with the intended audience – for example, focusing on the (negative) implications of climate policies for individual liberty and freedom. And as we argue in the next section, it is this level of communication and dialogue – speaking to the values and worldviews that people hold and weaving climate change into a story that sounds like it was written for them – that offers the potential to break the deadlock of public engagement with energy and climate change.

4.4 STORIES OF CHANGE: VALUES-BASED NARRATIVES FOR BUILDING PUBLIC ENGAGEMENT

We have argued that there are limits to the effectiveness of tweaking individual words and phrases to 'reframe' messages about climate change. But the limitations of this type of approach are not because language, words, and phrases are unimportant for public engagement with climate change. On the contrary, most attempts at linguistic reframing have not gone far enough, limiting themselves to the exchange of a small number of words in an otherwise fairly 'standard' message about climate change. In this section – and in keeping with a growing chorus of voices from across different disciplines and sectors – we argue for the importance of moving from simple alterations in message framing to a consideration of the role of narratives and stories as a way of building more meaningful engagement with energy and climate change.

The concept of using 'narratives' for communication has become increasingly common among climate communicators, and they are understood in different ways (Smith et al. 2014). There is no single, universal definition of a 'narrative', but they are usually thought to share some core features. Narratives have a setting, a plot (beginning, middle, and end), characters (heroes, villains, and victims), and a moral of the story (Jones and Song 2014). Narratives are a key method by which people make sense of the world, learn values, form beliefs, and give shape to their lives, and their importance as a communication tool is endorsed by scholars from across disciplines as diverse as linguistics, psychology, and literary theory (e.g. Herman 2013). In contrast to carefully controlled attempts to reframe climate messages, narratives are much more malleable, fluid, and non-linear, offering a mould into which facts, figures, ideas, and (crucially) values can be poured, rather than a fixed format for 'delivering' a message. As Simon Bushell and his colleagues put it (in a paper arguing for the importance of developing a strategic narrative on energy and climate change for the UK):

> [W]hile narratives can be constructed, planned, and promoted by specific actors to achieve desired objectives, they are not messages that get "delivered". They are not simply a "message" to be sent out to an audience in order to trigger certain expected (and predictable) behaviours, and 'they do not "spread" like viruses either. Instead, narratives can only be promoted by an actor; how they will be appropriated and interpreted by the audience and whether or not they will be retold and therefore continue to exist is merely something that the narrator can influence, not control...One... (cannot)...simply convey a narrative to a single audience – instead it will be commented on, interpreted, appropriated and retold by multiple actors, to multiple audiences. In this way the narratives take on a 'life-on-their-own' once they are put out into the public realm. (Bushell et al. 2016)

Echoing these sentiments, we view narratives not simply as 'longer' attempts at message framing: they are vehicles through which people can be connected (via the values they hold and the issues they care deeply about) to climate change and energy. There have been a number of demonstrations of the power of utilising values-based narratives for building public engagement. In one recent study with a large sample of the US participants, the same information about climate change (and the need to mitigate it) was presented in narrative formats that explicitly targeted groups with different cultural values (and then compared to the same information provided in a 'list' rather than a narrative – Jones and Song 2014). The study found that the information was more compelling when it was presented in a narrative that was 'congruent' with the cultural values of the group and that if the story did not line up with the group's cultural orientation, then it was no more effective than presenting the same facts in a 'list' format (Jones and Song 2014). Another US-based study explored the language and concepts around energy and climate change that might appeal to mainstream or 'swing' American voters found that value-orientated messaging was much stronger than a technical, scientific, or policy-oriented approach: people responded well to messages that linked energy and climate change to core American values such as leadership, exceptionalism, freedom, independence, and ingenuity (Western Strategies & Lake Research Partners 2009) – essential components of the 'American Story'.

This insight – that culturally 'congruent' stories about energy and climate change are at the centre of building public engagement – explains the findings of a raft of studies that have compared how people with different political orientations engage with different climate and energy 'solutions'.

As we discussed in the previous chapter, it is now well established that political conservatism predicts scepticism about climate change, especially in Anglophone nations (McCright and Dunlap 2011; Hornsey et al. 2016). Experimental research has found that framing climate change with reference to policies that are more congruent with conservative values (i.e. free market solutions to climate change) can reduce levels of scepticism (Kahan 2012; Nisbet 2009). One study termed this values-based engagement with different climate policies 'solution aversion', reporting that the US Republicans' opposition to the conclusions of climate science diminished when they were presented with climate policies and solutions that fitted more closely with their worldview (Campbell and Kay 2014). Put simply, people work backwards from 'answers' to climate change that they do or don't like and use this as the basis for evaluating the reality or seriousness of the underlying problem.

Importantly (and in the same way that a person's level of scientific knowledge is not straightforwardly related to their belief in climate change), favourability towards different climate policies is not driven by knowledge: there is no direct relationship between knowledge about a policy and support for it (Rhodes et al. 2014). In other words, you don't have to be a climate policy expert to be inspired by climate solutions that fit with your values, in the same way that it is not necessary to understand the physics of the greenhouse effect to be concerned about the impacts of climate change on the things you care about. More important than the actual solutions are the stories that grow around them and the meanings people attribute to different technologies and ideas.

Consider the contrasting ways that people respond to wind farms. Those who support wind energy equate the turbines with progress and preservation of the environment and feel reassured about the prospect of a clean energy future. And those who oppose them tell powerful tales about money-grabbing 'outsiders', defence of the landscape, and pledge solidarity in the face of undemocratic imposition on their community. Crucially, there is nothing written in the blades and motors of the turbines themselves that underpins these narratives. They are entirely social in nature. This means that telling the most powerful and compelling stories is the key – stories that relate to the aspects of people's lives they care passionately about (Corner and Roberts 2014). A climate solution is only as good as the story that surrounds it.

With this principle in mind, our organisation – Climate Outreach – has developed a programme of work focused on developing and testing a suite of different narratives on energy and climate change, all aimed at engaging citizens on the right of the political spectrum. In a series of publications, we have identified core beliefs that determine political conservatism (in the UK, but also in the wider European context) and have asked how different narratives could help build a bridge between communal conservative values and those that underpin engagement with climate change.

For example, in a 2012 report, we identified several narratives for communicating about climate change with centre-right citizens more effectively in the UK (Corner 2013a). As Fig. 4.1 shows, one of the narratives focused on making connections between the conservation of the 'green and pleasant land' and the risks that climate change poses to it (drawing on conservative values such as aesthetic beauty and responsibility), while another focused on 'safety and security' as a central theme:

In each case, we focused on mapping core aspects of conservative belief systems on to the energy and climate change discourse – attempting to find the overlap between the communal values of the centre-right and a societal response to climate change. One important consideration in constructing these narratives was to be conscious of the central lesson from the social psychology literature on values reviewed in the previous chapter: that messages about energy and climate change should be anchored, as consistently and firmly as possible, in communal values. As we explore in Chap. 5, focusing exclusively on the economic rationale for engaging with energy system change is unlikely to be an effective approach – for either side of the political spectrum.

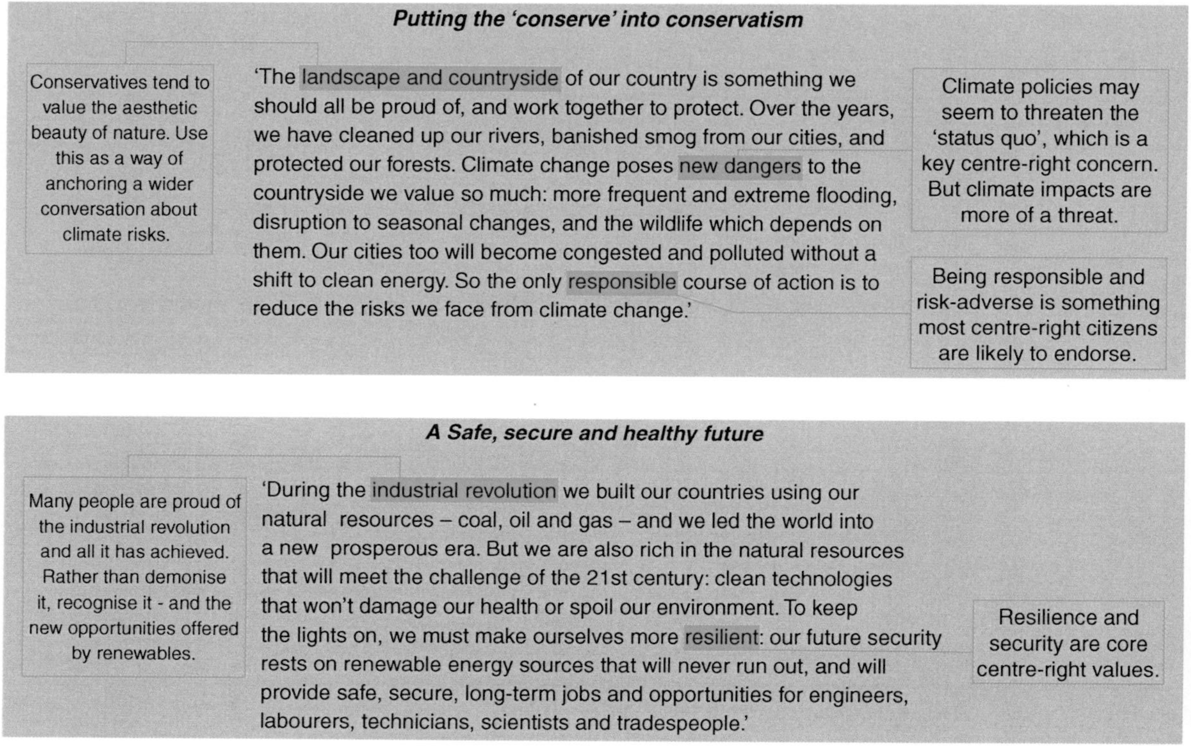

Putting the 'conserve' into conservatism

Conservatives tend to value the aesthetic beauty of nature. Use this as a way of anchoring a wider conversation about climate risks.

'The landscape and countryside of our country is something we should all be proud of, and work together to protect. Over the years, we have cleaned up our rivers, banished smog from our cities, and protected our forests. Climate change poses new dangers to the countryside we value so much: more frequent and extreme flooding, disruption to seasonal changes, and the wildlife which depends on them. Our cities too will become congested and polluted without a shift to clean energy. So the only responsible course of action is to reduce the risks we face from climate change.'

Climate policies may seem to threaten the 'status quo', which is a key centre-right concern. But climate impacts are more of a threat.

Being responsible and risk-adverse is something most centre-right citizens are likely to endorse.

A Safe, secure and healthy future

Many people are proud of the industrial revolution and all it has achieved. Rather than demonise it, recognise it - and the new opportunities offered by renewables.

'During the industrial revolution we built our countries using our natural resources – coal, oil and gas – and we led the world into a new prosperous era. But we are also rich in the natural resources that will meet the challenge of the 21st century: clean technologies that won't damage our health or spoil our environment. To keep the lights on, we must make ourselves more resilient: our future security rests on renewable energy sources that will never run out, and will provide safe, secure, long-term jobs and opportunities for engineers, labourers, technicians, scientists and tradespeople.'

Resilience and security are core centre-right values.

Fig. 4.1 Two narratives for engaging citizens with centre-right political values (Reproduced from Corner et al. 2015b)

In a subsequent report based on interviews with Members of the European Parliament (Marshall 2015), we argued that the concept of 'balance and realism' was central to the centre-right discourse on climate change in Europe and that speaking effectively to risk-averse and pragmatic centre-right values means framing climate policies in these terms. So while a standard campaign message asking people to support *more ambitious climate policy for a fairer world* might be effective for left-leaning supporters, the same message reframed for the centre-right would emphasise how climate policies could ensure greater stability (e. g. in the job market), reduce the risk of future threats (e.g. to beautiful landscapes through climate impacts), or secure a safer future for individuals and their families. None of these rationales are *alternatives* to talking about climate change; they are ways of beginning a conversation about the issue that are less likely to threaten or challenge the values of the audience. The central aim is therefore to expand and diversify the social reality of climate change – using language, values, and narratives that people with centre-right political views can engage with. Most recently, we tested a number of these different narratives in a series of structured discussion groups with members of the UK public who voted Conservative (Corner et al. 2016) and in a nationally representative online survey (Whitmarsh and Corner forthcoming). One of the themes we explored was the idea of 'avoiding waste' (as a goal of climate and energy policies and as a principle in keeping with conservative beliefs). Among the centre-right discussion groups, the notion of avoiding wastefulness was widely endorsed, with participants agreeing with sentiments such as:

> No-one likes to see things go to waste: it's just common sense. You teach your kids that it is irresponsible to waste things – to finish their dinner and not throw away food, and to turn off the lights in rooms when they're not using them. But millions of us live in old houses filled with gaps and holes that are drafty in the winter – we're literally throwing energy away. That's why energy efficiency is so important: who can argue with the idea of doing more with less?

In the survey, we compared a waste-themed narrative to a more 'standard' environmentalist narrative that talked about the need for urgent lifestyle changes to avoid 'climate chaos'. As expected, the narrative that focused on avoiding waste as the premise for engaging with energy and climate change was more popular among respondents to the right of the political spectrum. But among those on the left – who were more favourable towards climate change in any case – the waste-focused narrative was still rated positively. While the traditional environmentalist narrative produced the greatest discrepancy between left-leaning and right-leaning participants, ratings of the acceptability of the waste-focused narrative converged, suggesting that developing careful, evidence-based narratives for a particular audience need not come at the cost of excluding or disengaging others (Whitmarsh and Corner forthcoming).

While a good deal of our work at Climate Outreach has focused on engaging those on the right of the political spectrum – because of the continuing challenge of higher levels of scepticism among this group – we have also explored values-based narratives for engaging a range of other audiences too (including faith groups, as we discuss in Chap. 6). A final example which we describe here, though, is focused not on an audience defined by their political beliefs, but defined in terms of their national identity.

In what remains the only national-level attempt to systematically test the impact of narratives framed to appeal to a range of different values, Climate Outreach led a consortium of researchers in a project to produce a toolbox of language and narratives for public engagement on energy, climate change, and sustainable development more broadly in Wales (Marshall 2014c). In partnership with Welsh organisations and leading cultural specialists, key Welsh cultural values were identified and a series of trial narratives were tested in discussion groups.

In an early iteration of the Climate Outreach Narrative Workshop (which we describe in detail in the final chapter, offering it as a model for climate conversations), participants were first invited to explore their own attitudes, values, and concerns at length before commenting on the trial narratives (Marshall and Darnton 2012; Nash et al. 2012). Some of the core Welsh values that emerged as important were a sense of 'belonging' (linked to a shared national identity), pride in modest leadership (earned through hard work rather than through 'bragging' about being world leaders), fairness and 'fairplay', and a strong sense of attachment to the landscape of Wales (understood as a living, working space that includes all the people in it, rather than something of purely aesthetic beauty to be admired in a museum). The following narrative represents ideas and concepts that were popular across different discussion groups:

It's only fair that everyone should have rewarding and secure jobs and a decent standard of living that allows them to care for their families. But everyone knows that money and markets are not the only things that give people a good quality of life. In Wales we value the other kinds of wealth we possess in our relationships with our friends, family and communities.

The natural environment of Wales – our landscape, water, seas, air and everything that lives there – makes us passionate about Wales. This is a living and working landscape – not something to be put in a museum. There is not one part of Wales that hasn't been shaped by the hard work of people.

And there is another kind of environment that is just as important to people's quality of life. It starts at their front doors with everyday concerns: the condition of the pavements, vandalism and crime, litter, and the quality of the air they breathe.

It was our natural resources that built our country in the industrial revolution. And we are also rich in the natural resources that will meet the new challenges of climate change: the water, wind, forests and sun that can supply the energy needs of our people far into the future.

As we develop these natural resources we will hold onto the billions of pounds we send out of Wales for energy and can reinvest that in local jobs and opportunities for our own people.

This study represents one of the most ambitious attempts to test (on a national scale) narratives for engaging different audiences linked by a sense of national identity. This kind of approach – values-based narrative communication, derived from participatory public engagement – stands in stark contrast to most research on message framing. Our argument is that it permits a much deeper sense of public engagement to develop, where people (beyond the usual suspects) can hear a story about energy and climate change that sounds like it was written for them. While it goes without saying that any communication strategy – message, narrative, or story – should be honest and truthful, there is always more than one way to present information and ideas. And effective public engagement means telling a story about climate change that is both factually accurate and psychologically compelling.

4.5   CONCLUSION: NEW STORIES AND NARRATIVES FOR PUBLIC ENGAGEMENT

The approaches we have described in the preceding section are examples of public engagement that go beyond tweaking individual words or phrases and instead try to capture a deeper sense of what energy and climate change means to people's lives, their values, and the things they care about. Given the central role of values in determining how the public engages with energy and climate change, an important challenge is to identify ways of bridging the diverse values that any given group of individuals hold and the values that are congruent with a more sustainable society. By retaining energy and climate change at the centre, but weaving it into stories and narratives that connect with core communal values that people hold, values-based narratives offer an approach that can begin to expand the social reality of climate change.

As we discuss in the following chapters, the real benefit of using values-based language for public engagement comes through embedding the evidence base on narratives and stories in a dialogue-based, participatory context. If presented in a one-way process of 'messaging' to a particular audience, even a rich and detailed narrative approach is unlikely to produce much in the way of lasting persuasive impact. Stories about energy and climate change are things that people need to actively engage with, at their own pace, and on their own terms. They are not bite-sized chunks of information to be swallowed whole.

And, crucially, stories and narratives have the potential to take on a life of their own. Once a powerful narrative begins to spread, its growth can be exponential, the ideas it contains reaching far beyond the individual, group, or context where it was initiated. Through social networks – and via social media – stories can evolve, providing an opportunity for different groups and communities to develop a sense of ownership over climate change in a way that facts and figures will never engender. There are any number of practical barriers that stand in the way of decarbonisation, but mobilising creativity and imagination is also a crucial resource as societies collectively project and ponder a different type of future (Yusoff and Gabrys 2011).

What are the things that make people laugh, inspire them, or fill their conversations with friends? For most people, the answer will involve culture, not cognition. It follows that mobilising our cultural and creative resources might be as important for public engagement with climate change as technological or political changes. However, save for a few notable exceptions, there has been a gaping hole where creative energy should be.[6] Climate change theatre and films are thin on the ground. The situation is barely any different in the world of literature and storytelling. While there are a handful of examples of climate change-oriented novels, it does not seem to have fired the imagination of authors, and songs about climate change are notable by their absence.[7] That art provides a vehicle for bringing dry political sentiment or factual information to life is certainly not a new observation, but there is huge potential in harnessing the power of creativity to bring climate change narratives to fruition (Nurmis 2016).

Words, language, frames, and narratives (and visual tools that reflect the same principles) are absolutely central to building public engagement with energy and climate change. But even if people connect with a narrative about energy or climate change, does this tell us anything about their individual behaviours or other 'actions' they could take? Is it enough to engage with people's values, or do we also need an account of how rhetorical engagement becomes behavioural reality? In the next chapter, we explore the history of attempts at engaging individuals at the level of their personal behaviours. While a great deal of effort has gone into identifying ways of changing and shaping individual behaviours, we argue that much of it has been misplaced or misguided. But by seeing individual behavioural changes in the wider context of engagement with energy and climate change (incorporating social, cultural, and political shifts), we argue for a new, more integrated and holistic approach: shifting the focus of behaviour-change campaigns from 'nudge' to 'think'.

NOTES

1. http://www.yaleclimateconnections.org/2013/07/the-president-as-climate-change-communicator-how-obama-delivered-his-policies/. Accessed 23 June 2016.
2. http://www.energybillrevolution.org/. Accessed 23 June 2016.
3. http://www.wearefuterra.com/our-projects/sell-the-sizzle/. Accessed 23 June 2016.
4. https://www.thersa.org/discover/publications-and-articles/rsa-blogs/2015/12/paris-climate-agreement-the-good-the-bad-and-the-authentically-positive. Accessed 23 June 2016.
5. Roberts, D. (2016) Is it worth trying to 'reframe' climate change? Probably not, Vox Energy and Environment [Online], Available from: http://www.vox.com/2016/3/15/11232024/reframe-climate-change. Accessed 23 June 2016.

6. http://www.theguardian.com/sustainable-business/art-climate-change-communication; http://platformlondon.org/oil-the-arts/; http://www.capefarewell.com/latest/news.html; http://www.julies bicycle.com/. Accessed 23 June 2016.

7. http://crackmagazine.net/opinion/music/features/see-this-world-musi cians-climate-change/. Accessed 23 June 2016.

CHAPTER 5

A More Holistic Approach to Behaviour Change

Abstract Individual behaviours matter, but many early campaigns on energy and climate change trivialised the challenge by focusing on 'simple and painless' behaviours that had very little impact in terms of climate change. The principles of 'social marketing' and approaches such as the 'nudge' technique have grown in popularity. But while they are well suited to piecemeal behavioural changes, for a complex challenge like climate change, they are the wrong tools for the wrong job. To overcome the problem of 'rebound effects' and encourage 'spillover' between different behaviours, it is crucial to get beyond individual behaviours and engage at the level of values. The fourth principle is *moving from 'nudge' to 'think' as a strategy for public engagement*, promoting a sense of climate citizenship rather than following a prescriptive green lifestyle.

Keywords Behaviours · Social marketing · Nudge · Rebound · Spillover · Climate citizenship

© The Author(s) 2017
A. Corner, J. Clarke, *Talking Climate*,
DOI 10.1007/978-3-319-46744-3_5

PRINCIPLE 4: SHIFT FROM 'NUDGE' TO 'THINK' TO BUILD CLIMATE CITIZENSHIP

5.1   GIVING UP ON BEHAVIOUR CHANGE?

So far in this book, we have said very little about a central aspect of public engagement: people's individual behaviours, lifestyle choices, and the social practices that drive energy use (Whitmarsh et al. 2011; Spurling et al. 2013). In Chap. 1, we suggested that the easy distinction between technological changes, political or economic decision making, and individual attitudes and behaviours was in fact a false one – because human decision making underpins it all. But our analysis of building deeper, more committed public engagement with energy and climate change has so far avoided the question of whether – and how – the energy-consuming behaviours of individuals and communities have a role to play in a proportionate societal response to climate change.

This might seem a glaring omission; surely any strategy that stops short of promoting more sustainable behaviours and lifestyle choices at the individual level is not really a strategy for public engagement at all. After all, it is estimated that in the US, a 20 % cut in carbon emissions could in principle be obtained at the household level (Dietz et al. 2009a). This is not an insignificant level of reduction: the combined behavioural choices and social practices of individuals form a crucial piece of a proportionate societal response to climate change. International flights, meat and dairy-heavy diets, and high-mileage motoring all contribute significantly to climate change.

Many early attempts at public engagement on climate change in the 1990s and 2000s were characterised by a central focus on the role of individual behavioural changes – ordinary citizens 'doing their bit' through reducing their personal carbon footprints.[1] But over the past decade, there has been a concerted shift among campaigners, with many no longer talking about the role of individual behaviours at all. In part, this reflects a notable lack of success in terms of measurable outcomes of initiatives to change behaviours. Beyond some iconic and largely symbolic shifts (e.g. recycling and plastic bag reuse – both relatively peripheral in terms of carbon emissions and climate change), many impactful behaviours (such as aviation, meat-eating, and private car use) remain as popular as ever or are proliferating. Several academic reviews of the effectiveness of behavioural interventions have concluded that although it has been possible to bring about some limited reductions in personal and household emissions, the longevity and endurance of these changes are unclear, and they do not come close to the scale of change required to meet rapid decarbonisation targets (Capstick et al. 2015b). Nor do they necessarily justify the financial and political support they have received from campaigners and policy-makers.

So should we simply give up on behaviour change?

In this chapter, we review a range of different approaches to promoting 'lifestyle change' among the general public, which help to explain why we have not placed a more central focus on strategies for changing individual behaviours in this book (and why targeting individual behaviours in campaigns has largely fallen out of favour). Our argument is that individual behaviours (and the combined lifestyle choices and social practices of millions of ordinary citizens) *are* vitally important. But researchers and practitioners should take a more holistic and integrated view of the role of individual behaviours in building public engagement with energy and climate change (Henwood et al. 2015). This means acknowledging the limitations of top-down, message-based strategies for public engagement and putting a much greater focus on opportunities for participatory engagement (in person or convened online) and public dialogue. In these fora, individuals can *reflect* on the relationship between the actions that individuals can take and the bigger picture on energy and climate change.

One implication of viewing lifestyle changes in this way is that behaviours are simply one representation of an underlying commitment to the importance of climate change (and a proportionate response to it). What constitutes a 'proportionate response' will mean different things for different people. For affluent citizens (who in any case tend to have the highest carbon footprints) there will be a range of significant behavioural actions that they can (and should) take, in order to live in a way that is consistent with a deeper level of engagement with climate change. They might, for example, be able to pay more for certain services, adjust their lifestyles relatively easily, and invest in pro-environmental decisions that have upfront costs. For individuals who are less well-off (or constrained in other ways – for example by renting a property rather than owning one or through living in a rural location), the range of actions that are feasible (and reasonable) to take might be quite different.

This understanding means that there is not a 'ten step programme' for everyone to follow or a universal 'standard' against which an individual's carbon footprint should be assessed. In the same way that wealthy nations, historically responsible for a greater proportion of carbon emissions, have a 'common but differentiated' responsibility to cut carbon at the international level, so different individuals will have different levels of capacity to change their lifestyles. To take just one illuminating example, in the UK, the top 10 % of 'emitters' are responsible for close to half of all emissions, while the share of the bottom 10 % of emitters is closer to 1 % (Brand and Boardman, 2008). Much of this difference is underpinned by household income – illustrating clearly that 'changing behaviours' is itself a deeply political question, not easily disentangled from the wider socio-political context (Capstick et al. 2015b).

The aim of behaviour-change campaigns should therefore not be for everyone to live identical, low-carbon existences, but for individuals' behaviours to be as consistent as possible – within the constraints they face – with a deeper appreciation of what climate change means and how society should respond to it. Our argument is that lifestyle changes have a key role to play, but that the extent to which they matter will vary for different people, at different times. Much more than cajoling or 'tricking' people into different behavioural choices, behaviour change follows a process of reflection: changing the focus of behaviour change campaigns from 'nudge' to 'think' (Involve, 2010; John and Stocker 2010).

5.2   Individual Behaviours in Individual Boxes

Social marketing is the systematic application of marketing concepts and techniques to achieve specific behavioural goals relevant to the social good (Lazer and Kelley 1973). The term emerged in the early 1970s (inspired by a suggestion that social goods like brotherhood might be 'sold' like commercial goods – Wiebe 1952), and it has since been used to describe a wide range of programmes and projects aimed at pro-social behaviour change – especially in the health domain (Hastings 2007). Social marketing grew out of the realisation that simply providing information about a particular behaviour – the 'pamphlet approach' – was an ineffective way of bringing about behavioural changes. One of the central principles of the social marketing approach to behavioural change is that concrete behaviours – rather than general attitudes or beliefs – should be targeted.

Applying this logic to the sphere of promoting low-carbon behaviours (Peattie and Peattie 2009), the starting point of many campaigns to engage the public on climate change in the 1990s and 2000s was the idea that building public engagement, and catalysing a cascade of behavioural changes that would amount to a significant reduction in person carbon footprints, was best achieved by starting with simple, easy, convenient behaviours (Crompton 2010). Turning off lights, and not leaving appliances on standby, became the central focus of many behaviour-change campaigns. Typically, these 'simple and painless' actions also had very limited impacts in terms of their carbon emissions. But the assumption was that these initial changes would be the start of a process that would end in environmentally significant shifts in behaviours.

For example, reusing carrier bags has become something of an iconic 'sustainable behaviour'. But whatever else its benefits may be, the carbon impact of killing off plastic bags is negligible. Like all simple and painless behavioural changes, its value hangs on whether it acts as a catalyst for other, more impactful, activities or support for political changes, and here, the evidence is not encouraging. Academics at Cardiff University analysed the impact of the introduction of the carrier bag charge in Wales. Although their use reduced dramatically, rates of other low-carbon behaviours among the general public remained largely unaffected, with the authors concluding 'we do not believe that the increase in taking a bag when going shopping is a causal predictor of an increase in...other sustainable behaviours' (Thomas et al. 2016). To be clear: fewer plastic bags would be a small, good thing – but nudging, tweaking, or cajoling people into piecemeal behavioural changes like reusing plastic bags is not a proportionate response to climate change (and nor did it seem to be a very effective way of reducing plastic bag use until legislation was passed).

As it has become ever more clear that the positive cascade of behavioural changes envisaged by campaigners in the wake of interventions such as plastic bag-charging have not been forthcoming, critical voices on the value of this kind of approach have grown louder. As John Thøgersen and Tom Crompton put it, in a critique of the 'simple and painless' approach in 2009:

> The comfortable perception that global environmental challenges can be met through marginal lifestyle changes no longer bears scrutiny. The cumulative impact of large numbers of individuals making marginal improvements in their environmental impact will be a marginal collective improvement in environmental impact. [Thøgersen and Crompton 2009]

At the heart of the critique of the 'simple and painless' approach are two concepts, which are really two sides of the same coin: the idea of 'rebound' effects in energy use (in essence, whether the carbon savings generated by one energy-saving behaviour are offset, reduced, or eliminated entirely through increased energy use elsewhere – Chitnis et al. 2013) and the notion of 'positive spillover' between different pro-environmental behaviours (in essence, whether one pro-environmental behavioural change will lead to others – Thøgersen and Crompton 2009). A good example of a 'rebound' effect would be an individual deciding to 'treat themselves' to a foreign holiday with the money they had saved on their energy bill through insulation measures (also sometimes referred to as 'moral licensing' – Tiefenbeck et al. 2013). A demonstration of 'spillover' might involve someone applying behaviours practised in the workplace to their personal life (Rowson 2013).

The assumption that some kind of positive spillover will be set in motion is important because small, individual changes in energy-saving behaviour are in themselves insufficient to justify investing resources in. Similarly, when 'rebound' effects occur (typically indirectly by money saved on one low-carbon behaviour being spent on other activities that have a carbon implication), the assumed gain of the behavioural intervention is significantly reduced.

In reality, it is almost impossible to avoid rebound effects of some kind (Chitnis et al. 2013; Rowson 2013). In the absence of 'upstream' interventions like a carbon tax (in effect 'labelling' the carbon impact of all energy-using behaviour), any transaction (for services, goods, or even investment) has an invisible and unmarked carbon footprint. Rebound effects speak directly, therefore, to the need for a public-facing narrative on energy and climate change that places behavioural changes on the demand side in the context of a decarbonised energy supply and joins the dots between the many different initiatives that collectively define our societal response to climate change.

There are plenty of campaigns which appear to be completely 'oblivious to'? of the relationship between one pro-environmental behaviour and another. One striking and bizarre example saw consumers encouraged by a supermarket to 'turn lights into flights' by earning 'airmiles' through the purchase of energy-efficient light bulbs (Chitnis et al. 2013). It is a particularly nonsensical example, but the lesson is clear: if individual low-carbon behaviours are treated as compartmentalised and discrete, there is no reason to think that positive environmental actions will not be completely reversed by carbon-intensive activities elsewhere.

If the evidence on rebound effects suggests that the 'gains' of energy-saving behaviours are often not what they seem, then the evidence for the opposite effect – positive spillover between one behaviour and another – is similarly lacking (Austin et al. 2011). One US study found that compared to other equally 'environmentally aware' participants, people who more actively practised green consumer behaviours were also more likely to support wind energy (Thøgersen and Noblet 2012), but the empirical evidence is mostly correlational (showing that certain behaviours cluster together), not that one change leads to another (Thomas et al. 2016). Most experts caution against a simplistic understanding of spillover involving the conscious, sequential 'spread' of behaviours. Energy use is highly context dependent, and this poses serious barriers for spillover from one situation to another (e.g. from the household to the transport domain).

There are many situational barriers that prevent spillover and many systemic reasons why rebound effects occur (in effect, because every act has a 'carbon consequence'). However, these are most likely to be overcome if a clear, coherent narrative is in place that sets out how different behaviours relate to each other. Providing an opportunity to engage with and reflect on the 'big picture' is crucial. Without a clear sense of what climate change means for people's lives, or how the energy system is changing and why (and how different behavioural actions individuals could take relate to national-level policy), even the best-intentioned individuals are unlikely to be able to achieve much consistency in pro-environmental behaviours.

The problem – as with rebound effects – is that certain conditions must be fulfilled before a sequence of changes occurs (e.g. because of a desire to be consistent in behaviour; or due to increased self-identification with a sustainable lifestyle – Whitmarsh and O'Neill 2010). A small number of studies have asked what these conditions are. And these studies suggest that the *reasons* behind behaviours – and the values that differently framed messages speak to – really matter.

5.3   Save Money, Save the Planet?

Another key insight from the social marketing playbook is to start from 'where people are' rather than 'where you think someone should be'. In many ways, this is a sound principle, and understanding the audience of a campaign or communication is never a bad idea. But the logic unravels a little when applied to a challenge of the depth, and complexity of decarbonisation, as the following examples help to explain.

In an experiment with students in the UK (Evans et al. 2013), participants were encouraged to think about either the environmental or financial reasons for car sharing (or a combination of both types of reasons). In effect, the study participants were provided with messages that spoke to either communal or self-focused values, as reasons for engaging in a low-carbon behaviours. The researchers then recorded a series of subsequent energy-saving actions that participants could potentially engage in. Those who had been 'primed' with environmental reasons for car sharing were more likely to recycle their materials at the end of the experiment than participants who had been primed with either financial or a mixture of financial and environmental reasons for car sharing. Put simply, there was no financial benefit in recycling their materials, so no 'spillover' was observed. Similar results were obtained in a Dutch study comparing the effectiveness of messages based on different motivations for checking the tyre pressures on cars (Bolderdijk et al. 2013).

As the research on 'spillover' effects suggests, the reasons behind performing a particular behaviour matter. When a person acts for self-interested reasons, that person will perceive themselves as someone who does things for their own benefit. They will only engage in further behavioural changes if there is something in it for them – as soon as the 'sweeteners' dry up, so will their interest in sustainability. But if people begin to think of themselves as someone who cares about climate change, and who is invested – socially and culturally – in responding to climate risks, the chance that they will see the links between one behaviour and another is much greater. So the rationale behind promoting behavioural changes should not be 'because they will save you money'. This doesn't mean that saving money isn't important, or even essential, for many people as a principle that guides their choices. It doesn't mean that energy and climate policies should not make good economic sense. And there is no reason why policies to reduce energy use *should* come at a financial cost to ordinary citizens; if energy policies and behavioural changes can save people money, then so much the better.

But anyone seeking to promote low-carbon behaviours has to ask 'what happens next?' The logic of framing behavioural changes correctly is to position them as part of the bigger picture on energy and climate change, which appealing solely to people's wallets cannot do.

One recent example of a high-profile UK climate policy underscores the point. When the 'Green Deal' was announced in 2013, it was an ambitious programme, aimed at providing finance for household insulation and other energy-saving measures in millions of homes across the country.

The success of the Green Deal – as well as the many other policies that will follow over the coming years – hinges critically on people accepting the rationale for saving energy. The rationale is, of course, climate change and the need to radically reduce emissions from household heating. But this is difficult to tell from an announcement that launched the initiative:

> 'Energy saving has never been so attractive' – that's the message from Edward Davey today...Householders who use the Green Deal to make improvements such as loft insulation, solid wall insulation and new heating systems will qualify. Packages could be worth over £1,000. The more work households decide to have done, the more cash they could receive. To qualify for the Cashback Scheme, households need to book a Green Deal property assessment so they are then ready to have improvements installed under the Green Deal from 28 January and get their cash back.[2]

The full press release did not mention climate change once. Replace the term 'Green Deal' with the only slightly more generic 'Good Deal', and it would be difficult to know what was being promoted at all. No one is being encouraged to think about what climate change means or how different behaviours (around the home and when commuting, for example) might be related. No one is being encouraged to think about climate change or the underlying rationale for the policy at all. The exclusively economic framing of the government's flagship public engagement policy sends a clear message: people should take part in the Green Deal because it will be financially beneficial.

If all that was necessary to tackle climate change was making a few, unrelated, financially beneficial changes to things like cavity wall insulation, then the 'cashback' framing of the Green Deal would be a brilliant public engagement strategy. But given that what actually needs to happen is a little more challenging than this – involving major changes in how we travel, eat, heat our homes, consume, and work – this approach seems short-sighted.

The now-abandoned Green Deal has been a missed opportunity for the UK government to begin a positive national conversation about climate change. If children were taught that they would receive a pound coin every time they resisted physically hurting another child, they would not learn that hurting others was wrong; they would learn that restraining themselves was profitable. But it is precisely this logic that runs through major government initiatives such as the Green Deal.

It should not surprise us that bribing people into acting in an environ-mentally responsible way doesn't translate into meaningful engagement with climate change.

The typical repost to this argument from policy-makers charged with implementing ambitious schemes like the Green Deal, or campaigners frustrated at the lack of interest in their climate initiatives, is that climate change is simply too much of a 'toxic' brand to build a major initiative around. Knock on someone's door and talk to them about climate change (so the argument goes), and you will find the door rapidly closed in your face. Ask them if they would like to save money on their energy bills and the door is more likely to stay open. But this is an argument for working harder to expand the social meaning and ownership of climate change (so that climate change is no longer 'toxic' on the doorstep), not simply leaving it out of the conversation altogether.

It is precisely this logic that has contributed to a sustained process of social silence (Corner 2013b) around climate change: it is difficult to think about, and has become stigmatised; therefore, the best thing to do is to ignore it. But ignoring it simply kicks the challenge of building public engagement with climate change into the long grass. In the same way that people will not make the link between different individual behaviours unless there is an integrated and holistic narrative that binds the public discourse, framing climate policies in a narrow, economic way (and avoid-ing any mention of the reason they were developed in the first place) is a false economy. The longer we wait to initiate a meaningful conversation about energy and climate change, the harder it will be.

5.4   From 'Nudge' to 'Think': The Case for Building and Diversifying Climate Citizenship

Over the past two decades, a huge amount of time and effort has been expended trying to understand how to persuade, cajole, or regulate people into more sustainable patterns of behaviour. But in our eagerness to under-stand the drivers of behaviour, and our enthusiasm for measurable beha-vioural outcomes, we may have overlooked a critical point: that sustained and substantive behavioural transformations come not from gradually 'repro-gramming' our behaviour but from internalising the reasons for doing so.

Perhaps the most famous example of a school of thought that inten-tionally, explicitly avoids engaging with the underlying reasons for beha-vioural changes is the so-called nudge approach (Sunstein and Thaler, 2009).

The basic logic of the nudge approach is that by making small changes to the environment in which choices are made (known as the 'decision architecture'), behaviours can be 'nudged' in the right direction. It is a popular strand of the discipline known as behavioural economics, which acknowledges that people have biases and are not purely rational automatons. In this sense, the nudge approach is a progressive form of economics, and it has achieved a phenomenal level of policy capture, with 'nudge units' established in more than fifty nations around the world. Its self-styled blend of 'libertarian-paternalism' (Sunstein and Thaler, 2009) offers a tempting prospect to policy-makers: behavioural changes that don't require public-facing campaigns (or the budgets to support them).

A classic 'nudge' might involve changing the default setting that an individual encounters when faced with a choice. Several national governments have switched to an 'opt in' default on organ donation, which has vastly increased the number of people prepared to donate their organs after death (Thaler 2015). In a recent study of over 40,000 households in Germany (Ebeling and Lotz 2015), energy users were given the option of 'opting in' or 'opting out' of a green energy tariff on an online energy-supplier's web site (the green tariff which was slightly more expensive, but 100 % renewable). There was a significant jump in the percentage of people choosing the green energy tariff when the default was set to 'opt in': 6 %, compared to less than 1 % when they had to opt in themselves.

Similar to social-marketing approaches (which have produced notable successes in shifting health behaviours), the nudge approach can point to successes in changing easily compartmentalised individual behaviours, where decisions are largely a matter of personal choice and where the risks and benefits of behavioural choices are felt by the individual concerned. But interestingly, the authors of the energy tariff study interpreted their findings in the following way:

> Why are choices of 'green' energy particularly suitable for behavioural interventions using defaults? It is plausible that decisions that are highly relevant for one's moral identity are particularly influenced by default setting. As previous research has shown, individual morality is an important driver of pro-environmental behaviour. Actively negating one's moral convictions regarding the environment by opting out of a pre-selected pro-environmental option might be much more aversive compare with not opting in. Therefore, defaults could be particularly effective in the domain of environmental decision making, including energy choices. (Ebeling and Lotz 2015)

This raises the question of what is doing the 'work' here: is it the nudge or the deeper moral conviction? And where did that moral conviction come from in the first place? Our argument is that while nudge may be a step forward from the long-discredited 'rational actor' model of human behaviour (Sunstein and Thaler, 2009), it is a thin and ultimately inadequate form of public engagement for a challenge like climate change and decarbonisation, unless some more substantive social and psychological ideas are invoked (such as values, worldviews, or moral convictions).

Proponents of nudge-based approaches sometimes argue that because many (perhaps most) behaviours are habitual, expecting people to intentionally alter their decision making is unrealistic. But even the most ingrained habitual behaviours can be altered by making the behaviours 'conscious' and targeting the context in which they occur. When habits are disrupted by events/decisions (e.g. through relocation or a new job), behaviour-relevant information becomes more salient and influential, providing an important window of opportunity to intervene (Verplanken et al. 2008; Kurz et al. 2014).

Ultimately nudge is an 'unthinking' approach to behaviour change – when in fact precisely the opposite (a conscious process of reflection on the complex and multifaceted challenge of building a societal response to climate change) is required. In order to reflect on what holds the dozens of different behavioural choices that comprise a 'low-carbon lifestyle' together (and how these behaviours in turn are only one part of a suite of responses to climate change that encompass everything from energy infrastructure to 'divesting' from fossil fuel companies – Rowson and Corner 2015), a turn towards participatory public engagement, and away from 'nudge', is long overdue.

In a 2013 Climate Outreach report (Corner 2013b), we advocated for a programme of debates and conversations, begun not by green groups or climate campaigners, but by representatives of different communities (a concept we explore in detail in the next chapter). These events would be designed not to make an economic case, put forward scientific facts, or win an argument, but to allow people to reflect on what climate change (and a societal response to it) means. Isolated examples of these kinds of initiatives have taken place before (for example, the World Wide Views project, an exercise involving hundreds of people from around the world just prior to the UN climate change negotiations in Copenhagen in 2009, repeated again in a somewhat different format prior to the Paris UN meeting in 2015[3]). When they have occurred, a striking pattern has been observed: people move from disinterest to a position of engaged concern (Dietz et al. 2009b).

In our own work, developing precisely this kind of model for holding climate conversations in Scotland,[4] we have witnessed the very same transformation. These conversations are not aimed at reaching agreement on particular policies, and they do not focus on reducing participants' personal carbon footprints.[5] Social consensus is sought only to the extent that participants reflect on what climate change – and a proportionate societal response to it – means to them.

We do not wish to suggest that we have invented the idea of applying the principles of participatory engagement to energy and climate change. There is a rich history of using dialogue-based and participatory methods in research projects to study public attitudes towards the energy system (Parkhill et al. 2013), climate change (Hobson and Niemeyer 2012), and particular energy technologies (Devine-Wright 2007). And there have been occasional practical initiatives grounded in a participatory, conversation-based approach – for example a Scottish National Heritage project on climate change and local landscapes (Land Use Consultants 2011) or a UK-wide consultation on the subject of climate engineering technologies (Corner et al. 2012). And as we discussed in Chap. 2, there is converging evidence from public campaigns on other societal issues that participatory, peer-to-peer approaches are vital for developing meaningful engagement. But there has not yet been a coordinated, values-based programme of public engagement on climate change.

National-level (or even international-level) deliberation about the interconnected challenges of climate change and energy system transformation is not straightforward, but it is necessary. And where it has been carried out, it has been an effective means of anchoring wider social and cultural aspects of the climate challenge to scientific and technical discussions about energy system change (Pidgeon et al. 2014; Parkhill et al. 2013).[6]

At root, participatory public dialogues are about building a sense of 'citizenship' around climate change. 'Citizenship' is an easy word to throw around, although not as straightforward to define. But the basic concept of 'environmental citizenship' has existed for a number of years in the academic literature, and at the core of environmental citizenship is a recognition of *responsibilities* as central tenets (rather than *rights* as in liberal citizenship traditions – Dobson 2003). Environmental citizenship is based on a belief in the fairness of the distribution of environmental goods and in public participation in developing sustainability policy (Dobson 2010). In a recent review of the available evidence on environmental citizenship and pro-environmental behaviour, Andrew Dobson (2010) argued that if a sense of environmental citizenship can be fostered in individuals and communities, pro-environmental behaviour will be rooted in a commitment to the principles and values underlying it, rather than to financial or other types of external stimuli. This is a very different conceptualisation of the challenge of promoting pro-environmental behaviour to the social marketing philosophy of achieving piecemeal behaviour change using any method that in the short term 'works'.

But interestingly, although fostering environmental citizenship involves predicating specific behaviours on underlying values and principles, it is not necessarily 'the environment' that motivates environmental citizenship. Rather, it is a sense of fairness and justice between humans (requiring a commitment to conserving and protecting environmental resources) that plays the most important role (Dobson 2010). The notion of environmental citizenship therefore dovetails with the empirical evidence demonstrating that communal (rather than materialistic) values are more likely to produce pro-environmental behaviour (Crompton and Kasser 2009). In one study linking environmental citizenship with perceptions and practices around climate change, civic responsibility was found to be one of the most important motivations for participants responding behaviourally to climate change (Wolf et al. 2009).

Andrew Dobson (2010) argues that environmental citizenship can be fostered by increasing the opportunities for participation in local environmental decision making, by building and consolidating social capital and by working through existing agents of social change (i.e. social networks and community-based organisations). Clearly there is also a central role for education, which plays a crucial role in shaping the attitudes, values, and behaviours of children in later life. A survey by the Development Education Association (Hogg and Shah 2010) found that learning about 'global issues' such as poverty, sustainability, and trade creates agency around climate change in adulthood, reducing by half the proportion of people who feel that it is pointless to take personal action on climate change. Hogg and Shah (2010) reported that learning about climate change either in school or after school reduced the sense of powerlessness that an over-individualised presentation of climate change can bring from around 33 % (in those who had not taken part in similar learning experiences) to around 16 %.

Fostering a sense of citizenship among the general population is important if pro-environmental behavioural changes are to be embraced and maintained, with lifestyle changes seen holistically, as part of the bigger picture of climate change and energy policies. But climate citizenship need not be simply a form of 'environmental' citizenship. As we discuss in the next chapter, a key reason for the lack of widespread interest in climate change is that it has been strongly tagged as an 'environmental' issue in the public mind, to be dealt with by 'environmentalists'. The drivers of deeper engagement are not necessarily a concern for 'the environment' per se (as something external to human welfare), but in the fundamental principle of fairness and a responsibility to avoid harm (which could be applied equally to people or the planet – Howell 2014).

5.5   Conclusion: Scaling up Climate Citizenship

In this chapter we have argued that individual behaviours really do matter for public engagement with climate change, but as expressions of climate citizenship, rather than as ends in themselves. There is no 'one size fits all' green lifestyle or prescription for a low-carbon life, and the notion of behavioural changes is difficult to disentangle from wider socio-political factors such as people's income and capacity to control key energy-using areas of their lives. As is clear from the social psychology literature, even people with values that are congruent with pro-environmental behavioural changes can only adjust their lifestyles when the situation and circumstances permit.

Social marketing techniques and subtle nudges may be able to play a role in altering and shifting particular behaviours, but they must be anchored in the deeper notions of identity, values, and citizenship if they are to have a meaningful influence on promoting a proportional response to climate change – involving not just the widespread adoption of behavioural changes, but also the widespread acceptance of (or demand for) ambitious policy interventions (Ockwell et al. 2009). The notion of environmental citizenship implies a broader role for the public in engagement with climate change than that conceived by social marketing and behavioural economics – focusing not only on consumer behaviour but socio-political participation and civic engagement too (Brulle 2010; Dobson 2010; Hoppner and Whitmarsh 2010).

However, none of this means that behavioural changes do not matter, and the reason that (so far) widespread changes in impactful energy behaviours have not been forthcoming is because the wrong tools have been deployed for the wrong job. While social marketing can produce tangible changes in compartmentalised and piecemeal behaviours, the atomised approach it advocates (with no sense of how different behaviours relate to each other or how lifestyle changes fit with the bigger picture on energy and climate policy) are simply not fit for building a proportionate response to climate change. All the evidence suggests that if people have not taken on-board and internalised the reasons behind behavioural changes, they are unlikely to act in a consistently pro-environmental way.

Consistent behavioural change – as part of a more holistic understanding of the role of individual behaviours in responding to climate change – will only follow from a process of reflection, and this reflection protects against the risks of so-called rebound effects in energy use or the moral licensing of other high-carbon behaviours. Participatory public engagement is therefore crucial, and building a sense of climate citizenship (motivated by a whole range of different values, and using carefully chosen language as discussed in Chaps. 3 and 4) is key.

Importantly, climate citizenship need not be something that is the preserve of an educated elite who are comfortable with long, rambling discussions about abstract ideas, or a committed minority who wish to make drastic changes in their personal lifestyles.[7] While the full value of participatory engagement is achieved in direct, face-to-face interactions, there are now dozens of online platforms that facilitate genuinely interactive discussion and dialogue. Climate conversations can be inclusive, accessible, and grounded in diverse public values. But they will not happen without a much broader range of the public seeing climate change as something that is relevant to them.

So how can a much wider sense of climate citizenship be fostered? The next chapter offers an answer to this question: by promoting and nurturing new voices in the energy and climate change debate, overcoming negative social norms, diversifying the 'ownership' of climate change in the public mind, and building a new social reality around energy and climate change.

NOTES

1. http://www.theguardian.com/environment/2007/sep/13/ethicalliving.climatechange. Accessed 23 June 2016.
2. https://www.gov.uk/government/news/125m-green-deal-cashback-scheme-opens. Accessed 23 June 2016.
3. www.wwviews.org. Accessed 23 June 2016.
4. http://www.climatexchange.org.uk/reducing-emissions/climate-change-public-conversations-series/. Accessed 23 June 2016.
5. http://www.carbonconversations.org/. Accessed 23 June 2016.
6. www.wwviews.org. Accessed 23 June 2016.
7. www.carbonconversations.org. Accessed 23 June 2016.

CHAPTER 6

New Voices to Diversify the Climate Discourse

Abstract Building a new social reality around climate change means diversifying the imagery and spokespeople that represent climate change in the public mind. Nurturing and supporting representatives of diverse social groups is crucial – people who can speak with authenticity and integrity, using language and themes that lift climate change out of the 'green ghetto' in which it continues to reside. Harnessing the power of social norms is a way of catalysing individual behaviour changes, and participatory public engagement – climate conversations – should happen through existing social networks, and at scale, to have the greatest impact. The fifth principle for public engagement is to *promote new voices to reach beyond the usual suspects*.

Keywords New voices · Spokespeople · Messengers · Usual suspects · Green ghetto · Social norms

PRINCIPLE 5: PROMOTE NEW VOICES TO REACH BEYOND THE USUAL SUSPECTS

6.1   SEEING IS BELIEVING

If there is one image that represents climate change in the public mind, it is the polar bear (Leiserowitz 2006; Corner et al. 2015c). Repeated use of polar bear iconography by major NGOs like Greenpeace has provided a simple visual shorthand for the issue (Doyle 2011). But it has also reinforced the impression that climate change is a distant problem and arguably 'closed down' the climate discourse around a concept that is remote from people's day-to-day lives (Manzo 2010). Is our stunted visual vocabulary part of the reason that the social and cultural ownership around energy and climate change issues is still so narrow and restricted?

In a recent project called 'Climate Visuals',[1] we surveyed several thousand people in the UK, Germany, and the US. We found that polar bears (and other iconic images such as smokestacks) were quickly and easily recognised as signifying climate change (Corner et al. 2015c). They are effective ways of communicating to an audience that 'this story is about climate change'. But is it a story they want to hear? As well as an international survey, we carried out some in-depth discussion groups in the UK and Germany, and we include a snapshot of these comments here by way of illustrating the challenges faced in building a new social reality around energy and climate change.

When asked (before being shown any photographs) what image first came to mind when they thought of climate change, participants in our discussion groups readily made a series of associations – melting ice, a burning globe, fire, pollution, wind turbines, coal-fired power stations, and of course polar bears. And the findings of our survey were clear: these sorts of images were well understood and therefore more likely to be positively engaged with as a 'symbol' of climate change. However, these 'classic' climate images also prompted a significant amount of cynicism. One participant (in a comment characteristic of the UK and German groups) said,

> '... the polar bear ... makes me angry for some reason. Not because I'm like "oh no that's a pressing issue", but like "oh this is so annoying" I don't want to see them again. Like when you see a bad ad and you're like, oh leave me alone with that crap.'

A minority of participants were moved by images of polar bears. But even for these participants the effect seemed to be limited to the plight of the polar bear, rather than the climate issue as a whole. Others focused on the lessening impact of images that have been seen hundreds of times:

> 'It feels like, well I know it's been going on since I was in school, the late 80s, you see them in geography class, deforestation ...'

> '[These are] reproducing stereotypes that are already in my mind, also not really affecting me.'

While 'classic' climate images may be helpful for quickly and accurately 'symbolising' the issue of climate change for people who are unfamiliar with it, there is clearly a self-fulfilling risk in reusing the same imagery over and over again: it acts to 'close down' the climate story, instead of opening it up to new and diverse interpretations. Building a new social reality around energy and climate change means telling new stories that connect with the values of a much broader range of people. But while the move towards a more diverse and inclusive style of verbal and written climate communication has gathered pace, the iconography of climate change has remained relatively static.

As well as a limited visual vocabulary used to symbolise the meaning of climate change, there is also a very specific visual identity around the people who care about the issue. Another striking finding from our Climate Visuals project was the consistently negative attitudes participants held towards climate activists.

Images depicting protests (or protesters) attracted widespread cynicism and some of the lowest ratings in our survey. In our discussion groups, images of (what people described as) 'typical environmentalists' only really resonated with the small number of people who already considered themselves as activists and campaigners. Most people do not feel an affinity with climate change protesters, so images of protests may reinforce the idea that climate change is for 'them' rather than 'us'. Protest images involving people directly affected by climate impacts were seen as more authentic and therefore more compelling.

One image of a protester with his face painted blue, holding a 'Climate Justice Now' sign, was one of the most negatively received of all the photographs we tested. The individual was accused of being a 'frat guy' or alternatively someone who

> ...probably used the same face paint to paint himself at Glastonbury this weekend, and rubbed out climate and put Kanye West.

In a German discussion group, one participant objected to an image of a child at a climate change protest. The child, who was holding a banner in the shape of a foam finger, was described as:

> ...a classic example of jumping on the bandwagon. She wants you to take the threat seriously, but these balloons, and this foam finger, are the worst

for the environment. It's so outrageous, a lot of the time these protestors that are protesting climate change are doing things like this.

Overall, participants seemed tired of generic protest images. One picture prompted the comment 'For me, it feels like I've seen that image a 1000 times for pretty much every cause there is in the world.' On the other hand, specific campaign-related jargon – like 'divest' or 'climate justice' – meant little to the group members and mostly prompted confusion.

So why do most people feel such antagonism towards the standard iconography of climate change demonstrations? It is not straightforwardly attributable to scepticism about climate change. Analysis of the 'blue face' image revealed that the low ratings it attracted were not being driven by climate scepticism. People with high and low levels of scepticism were just as unlikely to feel motivated by the image. So it was not the case that this protest image was polarising: not even those concerned about climate change were particularly favourable towards it.

In our discussion groups, no one rejected the idea of human-induced climate change, but concerns appeared to be grounded in a dislike of the way the issue was discussed and communicated. One participant said climate change made him think of *'fascism'*; another told the story of when he got 'yelled at at a dinner party' for expressing doubts about it. Many of the participants expressed interest in and sympathy for social justice issues, and some concern about climate change. Most were not, however, sympathetic to environmentalists or images of environmental protest. When asked to say how they pictured environmental campaigners, one London group member suggested, 'either hipsters trying to be cool or...lunatic extremists'.

None of this should be taken to suggest that activists have done something wrong or deserve the vitriol that they sometimes attract. It is perhaps inevitable that activists who challenge, push against, or actively threaten the status quo will be viewed with a certain amount of suspicion by the mainstream. As we discussed in Chap. 2, activism (including radical activism) has been (and will continue to be) incredibly effective under certain circumstances. We have argued that there is valuable learning to be gleaned from some previous campaigns, but that in many important ways climate change does not straightforwardly 'fit the mould'. People simply don't see themselves in the polar bears, melting ice, or face-painted protesters. Unintentionally, the visual vocabulary that has developed around climate change has been sending out a clear message to the majority of people: this is not an issue that is relevant to you or something that you have a stake in.

Although their reputation may be undeserved, the people who are often the public face of energy and climate change campaigns – protesters and activists – are not perceived especially positively by the wider public. And the visual shorthand for the issue at a minimum does not do justice to the depth and richness of the climate change and energy debate. Partly because of the way that climate change has been portrayed by activists, it has an identity problem – specifically, that most people *don't* identify with it at all. Central, then, to the challenge of building a new and expanded social reality around energy and climate change is the basic process of cultivating public dialogue – breaking the pervasive social silence that surrounds climate change and harnessing the power of social norms and social networks to catalyse wider engagement beyond the usual suspects.

6.2   BREAKING THE CLIMATE SILENCE

Most people rarely talk about climate change, so there is a social silence around the issue, which acts as a negative social norm, discouraging others from engaging further even if they are inclined to do so (Corner 2013b). As we explored in Chap. 5, strategies like 'nudge' (creating the conditions under which certain decisions will be taken, but not explicitly encouraging people to change their attitudes or behaviours) contribute to this: people may be 'nudged' into pro-environmental behaviours, but they do not reflect on why this matters (or how it might relate to other aspects of their lives – Evans et al. 2013). Campaigns that are essentially motivated by climate change (the example discussed in Chap. 4 of the UK Government's Green Deal programme for home insulation is only one of many) often do not make explicit reference to it, for fear that it is a 'toxic' issue, of interest only to a narrow minority. But while it is true that people would probably find it easier to talk about the cost of wall insulation than what climate change means for society, the subject will remain the preserve of a narrow minority if it is studiously avoided in mainstream communication.

Some climate communicators counsel that the 'co-benefits' of climate change are what matters most (promoting the health or security implications of changes to the energy system), not climate change itself. But while connecting energy and climate change to people's interests and values through framing messages is a sound strategy, we have argued in Chap. 4 that this must be as a means of *starting a conversation about climate change*, rather than a substitute for discussing it at all. As we outlined in Chap. 1 (Leombruni 2015; Rowson 2013; Geiger and Swim 2016), the social silence around climate change has consequences. One recent study of American citizens concluded that:

> Because many Americans do not talk about climate change with even their closest friends or family, it is possible that some people do not know that those around them likely accept climate change. (Leombruni 2015)

Similarly, a poll of 2000 members of the UK public asked a number of questions that had a slightly unusual focus (ECIU 2014). As well as asking about people's own levels of knowledge and views regarding climate change and energy issues, it also included items which asked people to estimate others' views. For example, whereas surveys consistently show a large majority supporting technologies like solar and wind power, only 5 % of the ECIU survey thought that public support for renewables was between 75 % and 100 %. Most people surveyed (78 %) thought that up to half the population opposes renewables (when in fact, this number is much lower). In short, while most people in the UK support renewables, they think that most other people don't. Another study found that Australian citizens display a similar bias in their perceptions of climate change denial. Perhaps because of the over-representation of these views in the Australian media (relative to the actual proportion of scientists who dispute climate change), participants considerably overestimated the proportion of people who are sceptical about climate change in society (Leviston et al. 2012).

This type of misconception is known as 'pluralistic ignorance' in the social psychology literature (Shamir and Shamir 1997). When people misread a social norm in this way, it can lead them to suppress their own views, widening the divide between their own views and the views they attribute to others. This can further reinforce the false consensus and, at its most extreme, create a situation in which the majority of people keep silent because they fear that they are in the minority.

This misperception of the social consensus around renewables is reminiscent of the misperception of the scientific consensus on climate change.

But while a significant amount of research has focused on 'correcting' the misperception of the scientific consensus among the public (e.g. Lewandowsky et al. 2012; Maibach et al. 2014; van der Linden 2014), very little attention has been paid to the possibility that misperceptions of the social consensus around climate change might also play an important role. A social consensus could offer a more powerful metric for conveying agreement than a scientific one. How many of us would stand firm in our belief 'X' if 97 % of our friends, or those who we respected, thought 'Y'?

Recent work by the Common Cause Foundation (Common Cause 2016) underscores how important misperceptions of other people's values can be. In a representative study of the UK population in 2015, a standard values survey was conducted, asking people about the values they held. A clear majority (74 %) of respondents selected what the authors called 'compassionate' (i.e. communal) values, irrespective of their age, gender, or political persuasion. But when asked to answer the same questions about their fellow citizens – that is, when asked to estimate the values that other people hold – 77 % of respondents believed that other people held self-enhancing values. Put simply, while most of us value communal and compassionate values, we think other people are selfish in their outlook. And this misperception also has consequences: people who wrongly infer that others are selfish are less likely to be active citizens themselves, perpetuating the perception that no one else cares about investing in society.

If, as we have argued in this book, building a sense of climate citizenship is key to breaking the communication deadlock on energy and climate change, then the absence of accurate perceptions about what others think and value is a major barrier to overcome. In our own Narrative Workshop methodology (which we describe in detail in the next chapter), discussing and sharing values is a powerful way of revealing common ground. And because the silence around climate change is so pervasive, the act of spending dedicated time talking in a group about what climate change means for participants' lives is a radical departure from the norm. There is enormous value in pursuing strategies like this, which are the opposite of the 'nudge' approach. Getting people talking about climate change, taking on board the views of their peers, and updating their social misperceptions about others' views is crucial. Concluding her analysis of the critical role played by social networks in driving public engagement with climate change, Lisa Leombruni argues that there are important implications for campaigners:

the focus should be to get the public talking about climate change. Specifically, targeted campaigns to get climate acceptors to 'start the conversation' could significantly help raise awareness of the issue and make it an acceptable topic of conversation. Climate acceptors are in the majority: by reaching out to those around them – including family and friends, or especially those who are found to deny – acceptors could help start information cascades to help get climate change on people's agendas. (Leombruni 2015)

Breaking the social silence is crucial to build a broader and more inclusive sense of ownership and identity around energy and climate change. And as the next sections show, this process can be catalytic when it takes place among peer groups and social networks who share a common sense of identity or core values.

6.3   Social Norms and Social Networks

There are few influences more powerful than an individual's social network (Christakis and Fowler 2009) and the social norms that people are surrounded by. As we discussed in Chap. 2, health campaigns often target peer groups and existing social networks, in the hope that the spreading of positive behaviours will be more likely within groups of individuals who trust each other and pay attention to each other's behaviour (e.g. Abroms and Maibach 2008). The habits, beliefs, customs, and behaviours of people in any individual's extended social network – even if they do not know them directly or spend any time with them – have been shown to exert a powerful influence on people (Christakis and Fowler 2009). And there is a strong body of social psychological evidence that shows that social norms matter for energy and climate change too: in laboratory studies and more applied, practical settings, providing people with evidence of what others around them are doing has been shown to have a significant effect on attitudes and behaviours.

For example, Robert Cialdini (one of the leading proponents of social norms research) ran a series of studies showing that when hotel guests were informed that other people on their floor had reused their bath towels, they were more likely to reuse them too (Cialdini 2003). Academic research like this is now being put into practice by the energy company Opower,[2] who have used simple social norm strategies to achieve small but consistent savings on home energy use with their US customers.

Opower's approach is simple (in fact simple enough to be a 'nudge'!): every customer who receives an energy bill also receives information about how much energy they are using relative to their neighbours.

Social norms are a tried and tested method of influencing behaviour, but their effectiveness hinges on positive norms being available for promoting in the first place. For many sustainable behaviours, the problem is not that positive social norms aren't being highlighted, but that the norms are simply not there to promote.

Car use is a pertinent example: it is difficult to imagine how a campaign to reduce private car use could harness the power of social norms when the vast majority of people regularly choose this method of transportation. And even for behaviours where appeals to positive social norms are possible, they are likely to be drowned out by the torrent of messages promoting unsustainable behaviours from the advertising industry.

As a report by the Public Interest Research Centre and WWF argued, advertising has a doubly negative impact from the perspective of sustainability: it promotes values and beliefs that are antithetical to a pro-environmental self-identity (i.e. materialism), and it also seeks to increase the amount of products that are consumed (Alexander et al. 2011). So while positive environmental norms can be valuable tools for promoting sustainable behaviour, it is an uphill struggle to foster them in the first place. This highlights the need for more active methods of engagement than simply highlighting convenient statistics about other people's behaviour. As well as identifying instances where progress is already being made, making the case for behaviour change (actively engaging people beyond a social norm nudge) is also critical.

Social norm research has tended to be grouped along with social marketing and 'nudge' style approaches, and it is true that the basic notion of making visible other people's behaviour is consistent with a 'hands off' approach to engagement and behaviour change. In fact, Opower has worked directly with the UK government's Behavioural Insights team, the conduit for building 'nudge' approaches into UK government policies.[3] But social norms need not be a passive strategy for simply tweaking existing behaviours; they can be part of a much more ambitious approach (drawing on the cluster of approaches described in this book) that actively cultivates and supports a widening and broadening of the social reality of energy use and climate change. Social norm approaches have to be combined with more explicit and 'conscious' engagement strategies to be effective, ideally as part of a programme of participatory engagement where norms can not only be shared but reflected on and nurtured in new social networks beyond the usual suspects.

Unfortunately, despite the well-established understanding of the importance of social networks in general as an influence on individual behaviour, there is not much direct evidence about how social networks can be used to spread pro-environmental attitudes or behaviours. One recent paper – a rare example of a study that has applied a social-network approach to climate change – found that the stronger a person's social network, and the more they talked to friends and family about climate change, the stronger their level of belief (Leombruni 2015). The paper concluded:

> How you talk about climate change relates to the opinions you hold: it matters who you talk to, how frequently you talk to them, how close you are to them, and whether they share your beliefs or not. These interpersonal communication network behaviors and structures are predictive of whether individuals accept or deny climate change....(Leombruni 2015)

Evidence about social networks and the diffusion of behaviour in general (Christakis and Fowler 2009) suggests that pro-environmental behaviour change will be enhanced by targeting social networks (and the social relations between them) rather than individuals. Firstly, social networks are instrumental for creating a sense of social ownership (Rabinovich et al. 2010) around climate change. If engagement with energy and climate change is incorporated at this level (and becomes defining for a social group), it is an important way of combatting the over-individualisation of communication and engagement approaches. Engaging via social networks means thinking at the level of social identity, rather than in terms of compartmentalised and disjointed energy behaviours (which are often not even described as climate related in the first place, but are instead presented as economic choices).

Secondly, targeting engagement at social networks helps to enhance social capital – something that is critical for building the resilience to cope with and adapt to circumstances that threaten to deplete existing (psychological, social, or physical) resources (Rowson et al. 2010). Peer-to-peer learning also circumnavigates many of the problems associated with more 'top down' models of communication – not least that message-framing and language-based approaches are far more powerful when embedded in a dialogue. Engaging at the level of social networks not only helps to dispel misperceptions of others' views, but to reassure group members that becoming an active citizen around energy and climate change is not a wasted effort (because more people care than they might have realised).

And finally, focusing on social networks rather than isolated individuals holds out the prospect of harnessing the immense existing power, cohesion and influence of the hundreds of local, regional, national, and global networks which comprise our shared human experience – that is, public engagement 'at scale'. Of course, for the majority of people, their social network is unlikely to have climate change at its core. But social networks – Trade Unions, Football Clubs, Parent and Toddler groups – still perform a critical role in spreading change through society. Encouraging and supporting pre-existing social networks to take ownership of climate change (rather than a problem for 'green groups' to deal with) is vital – and in the following section, we describe a programme of Climate Outreach work where we have taken exactly this approach.

6.4   Keeping the Faith – Religion and Climate Change

Seventy-three per cent of people in the world identify with one of the five main religions: Christianity, Islam, Judaism, Buddhism, and Hinduism, the largest networks of shared identity in existence. Despite the growth of agnosticism in wealthier countries, these faiths continue to grow globally, especially in their more radical or evangelical forms. While climate rallies struggle to mobilise thousands for a single event, their vast networks bring together millions of people week after week in shared worship.

The approach to communication and engagement we advocate in this book foregrounds the importance of speaking to people's core values and identity. Our analysis of the fundamental role of values and identity – and the power of social networks – in creating a fresh approach to public engagement led us to think about faith groups as a potentially catalytic influence on climate change engagement. And in the lead-up to the UN climate conference in Paris, we were invited by the multi-faith GreenFaith network to develop and test narratives for mobilising people across Christianity, Judaism, Islam, Buddhism, and Hinduism, in order to inform their OurVoices[4] campaign.

We explored narratives around climate change based on metaphors and images that are found through the teachings of all the faith traditions, exploring polarities around cleanliness and pollution, sleep and wakefulness, light and dark. Some of our findings confirmed existing faith narratives around moral responsibility, respect for the natural world, and the need to express values through our actions (i.e. a motivation to avoid the value-action gap). However, we also found a strong resistance to overly moralistic and judgemental language that allocated blame or threatened punishment. In line with the evidence reviewed in Chap. 2 around the limitations and pitfalls of using fear and guilt to motivate climate engagement, even people of faith (for whom the notion of sin and redemption is familiar) found it difficult to incorporate climate change within this traditional moral framework. Many argued that contributing to climate change was not a straightforwardly sinful or immoral act and was more a product of ignorance or lapsed attention. We found that new language around the need to 'wake up' to climate change worked well across the faiths.

Although all faiths incorporate a strong environmental ethos, this is expressed through markedly different language and theology. The Abrahamic faiths – Judaism, Christianity, and Islam – believe in a creator God who has entrusted humanity with responsibility to be a caretaker or steward (Muslims prefer the word 'caliph') for creation. Hindus and Buddhists regard the natural world as part of a wider cosmological order within which humanity has a responsibility to respect its interdependence with other living things. We found that the only language that consistently worked across all faiths was the concept that the world is a precious 'gift' that requires respect. We also explored the concept of *balance*. We found that all faiths strongly endorsed the principle that there is a rightful balance to the world that is being unsettled by climate change.

This kind of research – grounded in participatory public engagement, among a group defined by shared core values, and with strong social ties and extended social networks – is a long way from decontextualised and individualised 'message testing' (in a psychology laboratory, or through an advertising campaign). The insights it produced are not 'magic words' that can be inserted into communications and expected to uproot the longstanding beliefs of someone who is sceptical about climate change or resistant to the dominant ideas at the heart of environmentalism. But they have the potential – if used by trusted messengers, who represent the social (or faith) groups they are engaging with – to produce a profound and catalytic shift in the social reality of climate change, moving it from the preserve of a narrow band of activists and into the mainstream.

6.5   TRUSTED VOICES: NEW SPOKESPEOPLE TO REACH BEYOND THE USUAL SUSPECTS

It may be a truism of communication, but it is one often overlooked by campaigners (who hope the importance of their message can overcome the disconnect between themselves and their intended audience) and researchers (who may recommend communication strategies without any sense of the real-world context in which they will be received and engaged with): the messenger matters as much, if not *more* than the message (Petty and Cacioppo 1984). Our argument is that it is crucial to nurture and support a range of trusted communicators to communicate energy and climate change messages. Who should speak for the climate?

Despite our critique of importing the trappings of formal science into public engagement on climate change, it is important to be clear that scientists are highly trusted voices in society, alongside GPs.[5] And climate scientists specifically are trusted voices on climate change for the majority of people (Pidgeon 2012). Indeed, research suggests that even though the public are more likely to trust a climate scientist if they believe their motive is to inform them of the consequences of climate change rather than to persuade them to take a particular course of action, they can be receptive to the latter if this is what they are expecting (Rabinovich et al. 2012). This underscores the high degree of public trust afforded to scientists as well as the importance of managing public expectations about the purposes of climate communication and engagement.

But most people are not able to engage with climate scientists on a regular basis – and many are unlikely to engage with them at all. Plus, many scientists are nervous or reticent about public engagement and especially the aspect of energy and climate debates that most people want to talk about: what does this mean for how we live our lives? Other non-scientific voices are also crucial to cultivate (especially people in the public eye), as a study of the US population shows. The sociologist Robert Brulle tracked public opinion on climate change over more than a decade, piecing together events and influences that had swayed views (Brulle 2012). Brulle's analysis pointed strongly to the importance of 'elite cues' – that is, signals and messages that people get from the media, politicians, and other high-profile voices. What they say matters – especially if they say nothing at all.

The evidence on whether 'celebrities' make for good messengers on climate change is mixed. Celebrities play a prominent role in many youth-focused advocacy campaigns, but their impact on public engagement is unclear, with some research cautioning that young people who follow celebrity culture are 'the least likely to be politically engaged' (Corner et al. 2015b). The limited amount of existing evidence suggests that the perceived popularity, credibility, and trustworthiness of a celebrity need to be considered carefully before involving them in climate campaigns. In one British study of 16- to 26-year old groups, some participants felt that celebrity involvement was a good way to raise the issue's profile but a greater number felt it was inappropriate due to their questionable legitimacy in terms of high-carbon lifestyles and relevant expertise (Corner et al. 2015c). Other studies have argued that the notion of 'eco-celebrity' (i.e. values-driven engagement with climate change, perceived as being central to the celebrity's public persona) can be very effective at mobilising young people in climate discourses and advocacy, including fans of the celebrity making climate-based connections with each other. A committed, consistent, and outspoken celebrity advocate such as Leonardo Di Caprio, for example, might fall into this category.

However, 'celebrity' is a relative term that depends on the audience being engaged. And perhaps more than the level of fame that a spokesperson on energy and climate change possesses is the congruence between their values and social identity, and that of the audience. One pertinent example is the former British Conservative Prime Minister Margaret Thatcher, renowned for her trenchant views on reducing the size of government, opposition to public ownership, and a general neoliberal economic outlook. Less well known is her early support for responding to the threats posed by climate change.

Thatcher had a background as a research scientist and fully accepted the science of climate change – in stark contrast to some in the British Conservative Party today who do not. Yet, as a 'conviction politician', she shaped climate change around the distinctly conservative values of duty, order, and prosperity. In 1989, at a time when public understanding of the issue was still limited and not yet polarised along party lines, Thatcher spoke of the dangers of climate change to the annual conference of her party (Marshall 2015). She anticipated that many in her party would be sceptical of 'left-wing environmentalism'. She openly mocked the left's dominance of the issue and asserted the superiority of conservative values. Throughout her speech she used words that would clearly signal conservative identity. In particular, she used language around the resonant conservative frame of the sanctity of 'life' – a word she repeats five times – and its opposites, depopulation and 'lifeless planets'.

In Fig. 6.1, an excerpt from this speech is adapted from a previous Climate Outreach report (Marshall 2015), annotated and with key conservative values in bold.

Thatcher's speech makes for a striking example precisely because of the polarisation that subsequently developed around energy and climate change in the UK (and elsewhere). But there have been other more recent examples of high-profile figures and institutions talking about climate change in their own language, using their own values and correspondingly engaging powerfully with their own audiences.

In 2015, in a Papal Encyclical[6] stretching to 42,000 words, Pope Francis set out new doctrine on climate change, covering science, politics, economics, and morality, and achieved something that thousands of climate activists struggle to do: focusing global media attention (temporarily) on climate change (Vulturis et al. 2016). This was not simply a technical retelling of the science of climate change or a meditation on the 'risks of dangerous climate change': this was a powerful and prescient call to arms, drawing as much on political passion as it did on scientific studies, and it resonated with a global audience of Catholics.

Other public figures and influential organisations speak to different audiences. The Governor of the Bank of England, Mark Carney, used a high-profile speech in 2015 at the insurers Lloyd's of London to warn that climate change will lead to financial crises, falling living standards, and 'stranded assets' unless the corporate world takes the risks of a changing climate more seriously.[7] By framing the problem as one of financial stability, Carney spoke to his audience in a language they could connect with and in terms they recognised. In a series of increasingly searing editorial statements during 2015 and 2016, the prestigious medical journal The Lancet has emphasised the inextricable link between a changing climate and human health, culminating in the launch of the UK Health Alliance on Climate Change, a coalition of major UK health institutions which raise awareness of the health risks posed by climate change.[8] From the Women's Institute,[9] to the National Trust, to the National Farmers Union,[10] other voices are emerging. But they must be nurtured, supported, and amplified in order for them to break through the climate silence and engage a much more diverse range of the public.

Mr President, when I spoke to the Royal Society about the environment *over a year ago, I spoke about the global threat of climate change. I set out the magnitude of the challenge we face.*

Scientific authority over the environment

We have to work to solve these problems on a **sound** *scientific basis so that our remedies will be* **effective**.

Rules create balance~~

It is no good proposing that we go back to some simple village life *and* halve our population *by some means which have not yet been revealed, as if that would solve our problems. Indeed, some of the Third World's* **primitive** *farming methods created the deserts and denuded the forests. And some of* Eastern Europe's **crude** technologies **polluted** *the skies and* **poisoned** *the rivers*

Asserting in-group superiority over environmentalists who oppose life and progress, and communists who destroy purity

It's **PROSPERITY** *which creates the* technology *that can keep the earth* **healthy.**

Economic narrative of increasing wealth through creativity

We are called conservatives *with good reason. We believe in* **conserving** *what is best-the* values *of our* **way of life***, the* **beauties** *of our* **countryside** *[that] have shaped our character as a nation.*

Assertion of conservative values; countryside and landscape and primary cultural 'values'

We have a special **responsibility** *not to let the towns* sprawl *into it.*

And to make Britain cleaner, we shall bring in a new Environment Bill to give us much **tougher** *controls on* pollution, litter and waste.

Cleanliness, pollution, litter, waste, 'sprawl' all damage the 'purity' of the countryside

Next month, I shall be going to the United Nations to set out our view on how the world should tackle climate change.

We have proposed a global convention – a sort of **good conduct** guide *to the environment for all the world's nations on problems like the greenhouse effect.*

'Good conduct' is conservative language of polite rule abiding

Britain has taken the **lead** *internationally and* we shall continue to do so.

Leadership

This is not only a question of acting **responsibly**, *though we do.*

Responsibility

Fig. 6.1   Former British Prime Minister Margaret Thatcher's speech to the Conservative Party Conference, 13 October 1989 (Adapted from Marshall 2015)

There is something deeper in us, an innate sense of belonging, *of sharing* **life** *in a world that we have not fully understood.*	A 'sacred value' and almost religious calling to value and preserve life
As Voyager 2, on its remarkable twelve year flight, raced through the solar system to Neptune and beyond, we were awe struck by the pictures it sent back of arid, lifeless *planets and moons.*	
They were a solemn reminder that our planet has the unique privilege of **life**.	
How much more that makes us aware of our **duty** *to* **safeguard** *our world.*	Custodians
The more we **master** *our environment, the more we must learn to* **serve** *it.*	Hierarchy and authority
That is the Conservative approach.	Re-assertion of in-group identity

Fig. 6.1   (continued)

The importance of nurturing and supporting trusted communicators – from a range of different sectors and demographics – is impossible to underestimate. Whether trusted peers in a participatory discussion, or value-congruent spokespeople in the public eye, cultivating new voices to speak for the climate is the crucial final principle for building a fresh approach to communicating climate change.

Notes

1. www.climatevisuals.org. Accessed 23 June 2016.
2. http://opower.com/. Accessed 23 June 2016.
3. http://www.behaviouralinsights.co.uk/. Accessed 23 June 2016.
4. http://ourvoices.net/. Accessed 23 June 2016.
5. https://www.ipsos-mori.com/researchpublications/researcharchive/3504/Politicians-trusted-less-than-estate-agents-bankers-and-journalists.aspx. Accessed 23 June 2016.
6. http://cafod.org.uk/content/download/25373/182331/file/papa-francesco_20150524_enciclica-laudato-si en.pdf. Accessed 23 June 2016.
7. http://www.bankofengland.co.uk/publications/Pages/speeches/2015/844.aspx. Accessed 23 June 2016.

8. http://www.thelancet.com/journals/lancet/article/PIIS0140-6736(16)30117-9/fulltext?rss=yes. Accessed 23 June 2016.

9. https://www.thewi.org.uk/campaigns/recent-campaigns-and-initiatives/women-and-climate-change. Accessed 23 June 2016.

10. http://www.nfuonline.com/cross-sector/environment/climate-change/. Accessed 23 June 2016.

CHAPTER 7

Five Principles and a Model for Public Engagement

Abstract This chapter summarises the five principles for public engagement described in previous chapters, proposes a tried-and-tested model for initiating a national climate change conversation (Narrative Workshops), and outlines criteria for judging whether a fresh approach was 'working'. New voices to catalyse engagement, new stories that resonate with diverse public values, and the cultivation of climate citizenship: these are the principles that can lift the energy and climate change discourse out of the margins and into the mainstream. It is time to start talking climate.

Keywords Five principles · Narrative Workshops · Public engagement · Fresh approach · Talking climate

7.1 FIVE PRINCIPLES FOR COMMUNICATING CLIMATE CHANGE

Despite two decades of awareness raising and campaigning, and an ever-growing academic literature on the subject, public engagement remains in a deadlock: climate change is a scientific but not yet a social reality. In this book we have proposed a fresh approach to public engagement, based on five core principles:

© The Author(s) 2017
A. Corner, J. Clarke, *Talking Climate*,
DOI 10.1007/978-3-319-46744-3_7

Principle 1: Learn Lessons from Previous Campaigns, and be Prepared to Test Assumptions

In Chap. 2, we asked, 'Is climate change different?' On the one hand, there are good reasons to think that climate change poses a challenge like no other. But there are important lessons to be learned from public engagement and campaigning on other thorny social issues. In this chapter, we described the pitfalls of relying on a fear-based approach to public engagement, the unintended difficulties that have arisen from framing climate change so strongly as an 'environmental' issue, and the importance of peer-to-peer engagement and maintaining a sense of social momentum. We argued that there is a key role for radical activism. But the limited progress made on public engagement is in part a failure of energy and climate change campaigns in building a wide and inclusive movement. Climate change remains socially and culturally associated with only a narrow band of activists, and this poses a barrier to making rapid and radical societal progress on decarbonisation.

Principle 2: Public Engagement Should Start from the 'values-up' not from the 'numbers-down'

However, the answer to this problem is not 'more science'. As well as being pigeonholed as an environmental issue, climate change communication has suffered because it has been dominated by technocratic targets and the 'big numbers' of the policy debate. In Chap. 3, we argued for approaching climate change from the other end of the telescope, building public engagement 'upwards' from the communal values that people from across the political spectrum hold, rather than 'downwards' from facts and figures about climate risks. Values are the starting point for public engagement – connecting energy and climate change to the diverse range of interests and aspirations that different people hold and helping to move climate change from a scientific to a social reality.

Principle 3: Tell new Stories to Shift Climate Change from a Scientific to a Social Reality

In Chap. 4, we argued that the way messages about climate change are framed matters – not because there are 'magic words' that can somehow transform someone's views, but because starting a conversation with people on terms they are comfortable with is the first step to building (and sustaining) their engagement. There are limits to the effectiveness of tweaking individual words and phrases to 'reframe' messages about climate change. But the limitations of this type of approach do not indicate that language is unimportant for public engagement with climate change. On the contrary, most attempts at linguistic reframing have not gone far enough, limiting themselves to the exchange of a small number of words in an otherwise fairly 'standard' message about climate change. We advocated for the importance of moving from simple alterations in message framing to a consideration of the role of narratives and stories as a way of building more meaningful engagement with energy and climate change, to be used in participatory public dialogues (not simply in slogans and advertising campaigns), as a vehicle for engaging with diverse public values.

Principle 4: Shift from 'nudge' to 'think' to Build Climate Citizenship

In the same way that tweaking individual words is likely to have a limited impact, we made the case in Chap. 5 that although it is possible to generate limited and piecemeal behavioural changes in low-carbon behaviours with conventional behaviour-change strategies, it is not possible to 'sell' climate change like a physical product, and no amount of 'nudging' can amount to a proportionate strategy for long-term public engagement. The act of talking about energy and climate change is radical and essential, but there is clearly a need to move from talking to 'action', and we argued that while there is no 'one size fits all' green lifestyle or prescription for a low-carbon life, individual behaviours (and how they relate to social identities and political choices) still matter.

The reason that (so far) widespread changes in significant energy-saving and low-carbon behaviours have not been forthcoming is because the wrong tools have been deployed for the wrong job. While social marketing strategies can produce tangible changes in compartmentalised and piecemeal behaviours, such an atomised approach (with no sense of how different behaviours relate to each other or how lifestyle changes fit with the bigger picture on energy and climate change) is simply not fit for building a proportionate response to climate change. The evidence suggests that if people have not taken on-board and internalised the reasons behind behavioural changes, they are unlikely to act in a consistently pro-environmental way. Until climate change means something more significant at the level of people's values and social identity, public engagement will remain stunted and fragile – so participatory engagement, and a space for reflection on the reasons behind behaviours, is crucial.

Principle 5: Promote New Voices to Reach Beyond the Usual Suspects

Instead of chipping away at individual behaviours one by one, we argued in Chaps. 5 and 6 that the focus of campaigners' attention should be on building a sense of climate citizenship and expanding the sense of social ownership around climate change. There are few influences more powerful than an individual's social network, and the social norms that they are surrounded by, and public engagement campaigns should harness the power of these networks to catalyse new voices in the energy and climate change debate.

Because building a sense of climate citizenship and wider social ownership is key to breaking the communication deadlock on energy and climate change, then the absence of accurate perceptions about what others think and value is a major barrier to overcome. The silence around climate change is so pervasive that the act of spending time talking about it is a radical departure from the norm. There is enormous value in pursuing strategies like this, which are the opposite of the 'nudge' approach to public engagement. Getting people talking about climate change, taking on board the views of their peers, and updating their social misperceptions about others' views are vitally important: catalysing and maintaining a vibrant public dialogue is an end in itself. But as we show in the final section of this book, there are many tangible measures that would demonstrate that 'talking climate' was working as a strategy for widening and deepening public engagement.

7.2   What If We Create a Better World That Not Enough People Wanted?

Faced with the complexity of improving communication and campaign strategies, there are those who argue we simply don't have the time for it. In an article for *Vox* magazine (first referred to in Chap. 4) titled 'Is it worth trying to reframe climate change? Probably not',[1] the well-respected environmental journalist David Roberts disputed the value of spending time or energy on the social meaning of climate change:

> ... (I)n terms of any large-scale, well-funded, concerted effort to change the way people think and talk about climate change? Meh. It's the kind of thing that forever appeals to funders, but it's a huge undertaking with dubious chances of success at a fairly late stage in the game. Climate change is what it is. The thing to do is just keep plugging away at it.

Certainly, the clock is ticking, and no amount of clever wordsmithing can (on its own) keep fossil fuels in the ground or make international carbon targets a reality. But the idea that '*climate change is what it is*' is perhaps the single biggest misconception among environmentalists and climate change activists. The challenge is very much not to simply keep *plugging away at it.* As years of heads being banged against brick walls can attest, climate change doesn't communicate itself, and perceptions of energy technologies are driven by a wide range of factors that have little if anything to do with the technologies themselves. The challenge is to step back, take stock of the things that determine how people engage with energy and climate change – and the central importance of values, narratives, personal reflection, and participatory engagement – and approach the challenge of climate change from the opposite end.

Climate change will mean different things to different people, but the important thing is that it means something at all. And while we may not have time for endlessly fiddling about with language or visual communication while global temperatures progress apace, we also do not have time for climate and energy policies that cannot be sustained because they are not built on a solid foundation of public opinion. A well-known cartoon,[2] frequently circulated among climate communicators, satirises opposition to climate policies that seem unarguably a good idea. An exasperated professor, presenting to an audience at a climate summit, lists the many reasons (including energy independence, healthy children, clean air and water) for embracing the low-carbon transition, as an audience member asks, 'What if it's a big hoax and we create a better world for nothing?'

The cartoon unintentionally raises another issue: what constitutes a 'better world' is something that people have different views on. Cleaner air and water may be an uncontroversial social good, but other aspects of climate policies are, and will continue to be disputed. The response to the satirical question in the cartoon might go something along the lines of 'What if we create climate policies that not enough people wanted, and then find we're unable to sustain them?'

Precisely because the challenge of climate change is so urgent, and the scale of the transformation of the energy system required is so dramatic, we cannot afford to waste time on strategies for public engagement that are untested, misconceived, or simply unpopular. As a society, we must get from 'A' (where we currently are) to 'B' (a decarbonised society, and a less than 2 degrees rise in global average temperatures), and we must do it quickly. But too many campaigns start with the end point 'B' and try to persuade people that this is something they should agree with. The approach advocated in this book is different: using participatory engagement – climate conversations grounded in communal values – as a catalyst for developing stories that provide a pathway for a range of publics to move their version of 'A' to the carbon-constrained 'B' that we know we must achieve in the end.

None of this means that progress cannot be made or that there is not a right and a wrong direction to be headed in. As George Marshall argues in his book *Why Our Brains Are Wired to Ignore Climate Change* (Marshall 2015), the notion that climate change is the 'perfect storm' uniquely capable of outwitting our psychological machinery or social and political systems is something of a self-fulfilling prophecy. Marshall argues that if the challenge of transforming our energy system in response to climate change is unusually complicated, it is because it is so multivalent (i.e. it is open to multiple meanings and interpretations). We project ourselves, our biases, and our expectations about the past, present, and future on to climate change and see solutions that fit with our perspective at the expense of others. This is why – to quote Mike Hulme's book of the same title – we disagree about climate change (Hulme 2009). But Marshall argues that the 'wickedness' of climate change is also an opportunity if we approach it in the right way: it lends itself to an almost unlimited numbers of stories being told about how to respond to it.

In a sense, while we may not be 'wired' to deal with a problem like climate change, the evidence reviewed in this book suggests that we are 'wired' in other ways that offer more hope: communal values are not the exclusive preserve of one political ideology, and there are dozens of powerful, passionate perspectives that can be nurtured and promoted on climate change, if space is provided for them to develop.

Precisely because this seems to be a challenge like no other, and precisely because there is no meaningful point at which the problem of climate change is 'solved', ongoing debate and deliberation is crucial, and disagreement (in the short term) is not necessarily a problem to be stamped out but could instead provide the fertile ground for genuine solutions to emerge from. As Amanda Machin argues in her book *Negotiating Climate Change*, disagreement is not a 'bug' in the system that should be eradicated (Machin 2013). Disagreement and contentiousness is the natural state of affairs for most subjects where differences in perspectives are grounded in deeply rooted values. Our argument in this book is that disagreements are not inherently bad things, and that developing authentic, culturally credible narratives about climate change that reach beyond the green bubble unavoidably means articulating visions that do not cohere and may even conflict (but agree on the importance of the issue).

Machin's analysis is important, because it challenges the dominant assumption underpinning many energy and climate change campaigns: that it is possible to 'get beyond the politics' and implement green solutions that somehow transcend political disagreement with a vision of the future that everyone endorses. Inevitably, this means being prepared for a 'battle' between competing ideas. Whatever else climate change will bring, it will not somehow smooth over differences in political ideologies, and there is no one single story that has monopoly on the route between here and a sustainable future. According to Machin:

> What it means to combat climate change depends upon who you are, where you are from, and where you would like to go … if we want to improve the chances of climate change rising up the political agenda we cannot demand that people 'see reason'. But we can acknowledge the environment in our perspectives and identifications and distinguish our identifications in distinct ways.

All of this means that the approach we advocate in this book can provide the best tools for starting a constructive conversation about energy and climate change, but it does not prescribe a fixed behavioural or energy-policy pathway. If we are serious about significantly expanding the social reality of climate change, then we have to accept that there is not one single green narrative that communicators are asking people to sign up to: communicators should instead attempt to talk about climate change in a way that engages a diversity of values and then use these values as building blocks for a variety of different energy and climate change stories. This is not an argument for removing politics from climate and energy campaigns, but it is an argument for not pegging public engagement to one particular political perspective or policy pathway.

Quite the opposite of being a 'drag' on otherwise streamlined climate policies, public engagement – conversations about energy and climate change – provides the momentum they require to have longevity and a defence against governments 'back-tracking' on their commitments. When the general population is actively engaged with climate policies and why they matter, moving away from existing commitments is made much more difficult. When citizens are not engaged, climate policies are much more vulnerable. Managing climate change is a project that will easily outlive the current contours of political opinion: it follows that building as broad-based and inclusive notion of what it means, why it matters, and what we should do about it is a top priority.

Next, we describe in detail our Narrative Workshop methodology as a tried-and-tested participatory model for building engagement with climate change. If applied at scale and through existing social networks, this type of approach would significantly widen and deepen public engagement. But how would we know it had worked? In the last part of this book, we offer a definition of what meaningful public engagement with climate change would look like (and some suggestions for how to measure it).

7.3   A TRIED-AND-TESTED MODEL FOR CLIMATE CONVERSATIONS

Throughout this book we have argued for the importance of moving away from social marketing and nudge-style approaches, and towards participatory public engagement. Crucially, processes for public engagement should begin with people's values, and the Climate Outreach Narrative Workshop methodology provides a model for just such an approach (Corner and Roberts 2014; Shaw and Corner 2016). Developed from well-established principles of participatory engagement, and first formally trialled in 2011/2012 in work conducted on behalf of the Welsh Government (Marshall and Darnton 2012; Nash et al. 2012), Narrative Workshops aim to unearth the values and principles on which different people base their views about the world and build a bridge – a meaningful storyline – from there to a proportionate societal response to climate change. We have now run dozens of these workshops, working with different partners (including the Scottish Government and leading NGOs such as WWF) and with different target audiences (including centre-right citizens, faith communities, social housing tenants, and young people).

Our 'funnel' design starts with discussion of participants' values, concerns, and aspirations. We find this process typically leads to a recognition that there is a set of communal values held in common. The next step is to take the conclusions from this discussion and focus in on a more personal and localised interpretation of these general ideas: are these values common in the local community, and are they undergoing change? This conversation around change serves as a bridge into discussions about fears and hopes for the future. The conclusions emerging out of these conversations provide a 'lens' through which to discuss climate change and explore different language and narratives for public engagement.

The Narrative Workshop methodology relies on an informal approach to conversations, hence minimum use of slides and presentations. Instead we largely rely on participants' own voices to build a narrative arc through the workshop. This is designed to be an inclusive and accessible approach, as we draw on the values and beliefs that participants hold in common. Based on feedback collected from participants, people find this process has a significant, and positive, impact on their level of engagement with and concern about climate change, as the conversation develops. It is an approach that begins with people, rather than a particular policy proposal or technological response. As the wider research discussed in previous chapters shows, people are more likely to engage positively to climate change messages when they are presented within narratives that validate their values and identity, and the participatory context permits the full value of values-based framing and language to be realised.

Recently, we designed and tested a model – based on our Narrative Workshop methodology – for holding 'Climate Conversations' at a national level in Scotland, on behalf of the Scottish Government. In 2009, the Scottish Parliament unanimously passed the most ambitious climate change legislation anywhere in the world.[3] The Scottish Government recognises that it faces tremendous challenges in delivering on these ambitions and that success is dependent on the support and involvement of the Scottish public. Generating an ongoing and self-sustaining national conversation about climate change in Scotland will be an essential step in building that support. This pioneering project is the first of its kind in the UK and possibly anywhere in the world.

What is novel about our work in Scotland is not the idea of talking about climate change in a group. Approaches such as 'Carbon Conversations',[4] 'Carbon Ration Action Groups', and 'eco-teams' have all shown promise as a way of producing durable pro-environmental behaviour change (Capstick and Lewis 2008; Nye and Burgess 2008). Typically, people find their structured yet social nature to be appealing, with participants identifying a sense of mutual learning and support as a key reason for making and maintaining changes in behaviour. However, with a handful of exceptions, these group-based methods of engagement have focused exclusively on reducing the individual carbon footprints of group members and so (almost by definition) have overwhelmingly attracted participants who are already to some extent committed to the issue. In contrast, our Narrative Workshops – and the Scottish Climate Conversations model – are aimed at people with no prior interest or engagement in climate change.

The full potential of values-based, participatory engagement would be realised if it were focused on pre-existing social networks, who have strong ties but do not necessarily have climate change or energy at their centre. This could mean face-to-face interaction (and our experience suggests that participatory engagement is most powerful in this format). But it is not feasible to hold in-person climate conversations with millions of people – and certainly not in the context of a rapidly changing climate when time is of the essence.

However, a rapid and radical intervention in the public discourse on climate change need not be limited to physical discussion. There are an abundance of widely used and cost-free online platforms on which millions of people interact on a daily basis. Of course, these platform have been used repeatedly for *campaigning* on climate change. But they have rarely been used for climate conversations. The potential reach of a dialogue initiated by (for example) a major sports team through its social media channels, or a forum like Mumsnet,[5] would be considerable. By diversifying the voices talking about climate change, the social reality of the subject would be fundamentally altered. And in this space, a truly proportionate response to climate change could flourish.

Resilient climate solutions (whether individual behavioural changes or energy policies) require robust, resilient public support. Without it, progress on climate change is a hostage to fortune, vulnerable to sudden shifts in the political winds, or external shocks such as an economic downturn. So public dialogues – climate conversations – are not a waste of valuable time. On the contrary, they are the glue that can hold the diverse elements of a proportionate societal response together. We do not gain time against the ticking climate clock by abandoning or avoiding public engagement; we lose it.

In this book, we have argued for a fresh approach to public engagement that could deliver the kind of broad-based (and more substantive) support required to make meaningful progress on climate change. But how would we know that this alternative type of approach had 'worked'?

7.4   WHAT DOES MEANINGFUL PUBLIC ENGAGEMENT MEAN IN A POST-PARIS WORLD?

Without a doubt, debate and disagreement on its own is not enough to protect us from climate impacts or underpin the transition to a very different type of energy system. Conversations must have a basis in the reality of a changing climate. So it is crucial to have a basic framework within which climate conversations can operate.

The agreement reached in Paris at the UN negotiations in 2015 provides a framework with which to evaluate the progress of individual nations, because all countries have endorsed the principle of limiting global warming to less than 2 degrees above pre-industrial average temperatures. It tells us virtually nothing about how this abstract target should be achieved, but it offers a line in the sand, which other perspectives and positions can be anchored around. The challenge for the international community following the Paris agreement is not to set off on a UN-prescribed pathway to solve climate change, but to develop (at a national and regional level) plans, roadmaps, and policies which operate within the 2 degrees parameter.

Almost everything is still to be decided, and certainly the heavy lifting is ahead of us. But it does mean that the perfect time for opening a wide-reaching public dialogue on energy and climate change is now, and the fact that the UN's climate change framework commits all member states to doing precisely this is a positive step in the right direction.[6] As national plans develop and are implemented, they must be grounded in a foundation of public engagement. And the 'line in the sand' provided by the Paris agreement for national legislators has a kind of domino effect on how individual nations can approach and gauge progress on public engagement.

Because there is now an international yardstick, national plans must add up to 'no more than 2 degrees'. The challenge for engaging the public at a national level is to reach a point at which a solid majority of the voting population is not only supportive of, but actively positive towards, a national societal response that collectively meets (or ideally surpasses) national carbon targets. In this sense, there is a cascade effect from the Paris agreement, in that public engagement programmes can be tethered to something tangible and globally recognised. If, following five years of climate conversations and public engagement, the population of a given nation was no more supportive of a set of national policies that equated to their piece of the '2 degrees' puzzle, then the programme would be a failure. But if the level of public engagement had widened, and deepened, and committed public support for a proportionate societal response had increased, the programme would be a success.

The journey that individuals and communities will take mirrors that of national legislators. In the same way that the UN did not prescribe national decarbonisation plans, there is no single, prescriptive lifestyle or energy policy that is implied by nations seeking to implement a proportionate societal response to climate change. But there is (for now) a non-negotiable limit, at a global level, which all nations have endorsed and agreed to play their part in achieving. And this means that campaigners and communicators can judge progress in public engagement by comparing public attitudes, behaviours, and preferences against national policies and carbon budgets. For the first time, there can a reasonably clear metric against which to gauge public opinion: are most people (in a given nation) on board with a way of life and a set of policy choices that will deliver or exceed national climate targets?

This is not a question that it would make sense to initiate a public conversation with. As we have argued in Chap. 3, '2 degrees' (or any other abstract, technocratic framing) does not offer a useful public-facing way of starting a conversation. But researchers, communicators, and campaigners can use it as way of measuring progress in public engagement. The UK offers a good example in this regard, as it has well-defined targets for reducing carbon, which correspond (to some extent) to the ambition of the Paris agreement.

Researchers at Cardiff University took advantage of the relatively well-defined state of the UK climate targets, and used a tool which allows people to change the 'levers' of energy policy, in order to meet targets for 2050. Using a tool developed by the UK Government's Department for Energy and Climate Change (DECC), the Cardiff researchers asked participants to explore the 'My2050'[7] online tool which allows people to vary different aspects of the energy system using simple sliders, in order to meet the UK's targets for reductions in carbon emissions. In a paper evaluating the approach they used, they argued that:

Although the tool sets a number of constraints on the choices people can make, the data provided valuable insights into people's views and choices about a desirable future when considering multiple options and tradeoffs in the context of each other... Accordingly, and in contrast to offering scenarios constructed by experts, this gave participants the opportunity to develop their own scenarios reflecting their values and views on how energy systems should change. (Pidgeon et al. 2014)

While this example is obviously specific to the UK context, it gives a good sense of the type of structured exercise that can provide a focus for participatory engagement. If implemented at a national scale, on a regular basis, getting people talking about climate change, weighing up energy policy options, reflecting on what different scenarios and choices would mean for their lives and the things they value, would be a radical new approach to public engagement.

The focus need not stop at simply *accepting* existing climate targets; engaged citizens might well argue that they are inadequate (Shaw 2015). Indeed, a central claim in this book is that public engagement is not a means to an end, but an end in itself: the very act of promoting and maintaining a dynamic public dialogue will make climate policies more robust and more resilient. But if the new approach that we advocate to public engagement is to gain traction among campaigners and communicators, it will need to be synchronised with existing policy frameworks. The post-Paris national commitments are by no means perfect, but they are – currently – the best we have.

So what would meaningful progress (with individuals and communities shifting towards a position that reflects the realities of national and international carbon commitments) look like, in terms of attitudes, behaviours, and policy preferences? We argue that it would include (but not be limited to) the following three broad criteria.

1. Ownership and Identity

A key argument throughout this book has been the importance of expanding the social reality of climate change. We have suggested that by pursuing values-based strategies of participatory engagement, and using language that is explicitly designed to resonate with diverse public values, a much greater sense of social ownership and identity around climate change can be fostered.

Climate change is not only or even mostly about 'the environment'. But the tropes and cultural contours of environmentalism have defined climate change in the public mind. Catalysing new voices means widening the social reality of climate change. The challenge is not to 'turn' the vast array of un-engaged social groups into environmentalists, but to build new senses of ownership among these disparate groups. Narratives – whether verbal or visual – are the key means by which to achieve this. Participatory public engagement is the channel through which they should be promoted and applied, with trusted values-congruent communicators acting as catalysts for others in their social network.

A wider sense of social ownership and identity around the climate change and energy debate, and a sense of climate citizenship would manifest in the following ways:

- An increase in the relative importance of climate change compared to other social issues
- More regular social interactions around climate change, breaking the social silence
- Less negative views of environmentalists and environmentalism
- Higher visibility of climate change in popular culture, and innovative creative and cultural responses to climate change
- Prominent 'unusual suspects' talking about why climate change matters to them

2. Changes in Individual-Level Attitudes and Behaviours

We have argued that the focus of many early campaigns on 'simple and painless' behavioural actions was misplaced, that the challenges of behavioural 'rebounds' are significant, and that the elusive idea of 'spillover' between different pro-environmental behaviours is conditional on people engaging at a much deeper level than 'nudges' to their behaviour or via social marketing campaigns. However, there is clearly still an important role for individual attitudes and behaviours, albeit as part of a more complex system of influences that includes habitual, ingrained social practices, and a range of situational and structural constraints on individuals' lives (Whitmarsh et al. 2011). If behaviour-change campaigns are built around engaging with communal values, then a range of behavioural changes are more likely to follow. Put simply, behavioural changes follow a process of building climate citizenship, not the other way around.

However, that doesn't mean there is a single prescription for a 'green lifestyle' that everyone should follow. Behaviours are simply one representation of an underlying commitment to the importance of climate change (and a proportionate response to it). What constitutes a 'proportionate response' will mean different things for different people. For affluent citizens (who in any case tend to have the highest carbon footprints) there will be a range of significant behavioural actions that they can (and should) take, in order to live in a way that is consistent with a deeper level of engagement with climate change. They might, for example, be able to pay more for certain services, adjust their lifestyles relatively easily, and invest in pro-environmental decisions that have upfront costs. For individuals who are less well off (or constrained in other ways – for example by renting a property rather than owning one, or through living in a rural location), the range of actions that are feasible (and reasonable) to take might be quite different. In the same way that wealthy nations, historically responsible for a greater proportion of carbon emissions, have a 'common but differentiated' responsibility for cutting carbon at the international level, so different individuals will have different levels of capacity to change their lifestyles (Capstick et al. 2015b). However, a significant shift in public engagement would be likely to include the following aspects:

- An acknowledgement (rather than dismissal) of the role of individual behaviours in responding to climate change, and a willingness to adopt lifestyle changes as part of societal response to climate change
- Shifting social norms around pro-environmental behaviours; low-carbon lifestyles becoming aspirational and mainstream
- A reduction in 'rebound' effects and an increase in 'spillover' (i.e. a consistency across different types of pro-environmental behaviour)

3. Policy Preferences

The third category in which progress in public engagement should be visible and measurable is in aggregate policy preferences. To take the

UK as an example, there is generally strong support (in principle) for renewable energy technologies, as captured in national surveys (e.g., Parkhill et al. 2013). But there has been repeated opposition at a local level to the siting of renewable energy technologies. Progress on public engagement in terms of policy preferences would mean a solid and committed level of public support for a suite of changes in the energy system that correspond to national carbon targets. Committed public support would mean that it could not easily be blown around by the political winds (i.e. a change of government), and in fact, that wider and deeper engagement on energy and climate change would act as a 'backstop' against governments reneging on their existing or future commitments. Overcoming the so-called governance trap (Pidgeon 2012) on energy and climate change (where politicians won't push ahead of where they perceive public opinion to be) means starting with the public and building a chorus of voices in support of a progressive energy policy, not just lobbying politicians.

Having a highly visible, national (and multi-partisan) narrative is crucial to join the dots between the different elements of the energy system, as well as how these structural level changes relate to individual actions and behaviours. Progress on policy preferences would be likely to include the following:

- Widespread support across the political spectrum for changes in energy system (supply and demand) to meet carbon budgets
- An increasing number of people who recognise the trade-offs in energy system choices
- Strong individual visions of a low-carbon society taking a central place in the manifestos of all major political parties
- A consistent and coherent narrative from a cross-party coalition that extends beyond the electoral cycle

These basic elements allow participatory public engagement to be anchored in some very tangible national and international policy goals. They are both an alternative set of metrics for climate change communication researchers, and potential targets for practitioners and campaigners to aim for. Creating a much closer calibration between the worlds of research and practice on climate change communication would significantly increase the chance of public engagement progressing.

7.5   FINAL REFLECTIONS: NEW INFRASTRUCTURE FOR TALKING CLIMATE

One of the core ideas in this book is the importance of shifting climate change from a scientific to a social reality, so that committed public support for a proportionate societal response can take root across the political spectrum. A relentless focus on the science of climate change has swamped the climate and energy discourse, displacing other equally important dimensions of the issue: economics, culture, and psychological or social dynamics (Rowson and Corner 2015). And it has provided a stilted vocabulary with which to describe and discuss the issue that will define the twenty-first century. But the challenge is not to diminish the scientific foundations of the climate change and energy debate or bypass the trusted expertise of scientists. Instead, the science needs to be brought to life and to sit alongside the other crucial dimensions of the issue. And for this, new infrastructure for public engagement is required (Corner and Groves 2014; Corner and van Eck 2014).

We have advocated for conversations as the key channel by which public engagement can be widened and deepened. As we discussed in Chap. 2, emerging ideas around 'deep canvassing' have produced remarkable shifts in attitudes towards transgender issues in the US, prompted by reflective and considered doorstep interactions. Talking to people in this way about climate change would be possible, but even more powerful would be conversations within social networks, among people who already trust and respect each other.

But the right infrastructure does not yet exist to allow this to happen. Governments can sponsor or support public engagement activities but will always be viewed with a certain degree of suspicion by the electorate. Scientists certainly have a role to play, but they do not (for the most part) speak a language on energy and climate change that resonates beyond the ivory tower. Campaigners may have social and cultural currency, but they do not have scientific expertise and therefore may find themselves challenged in a different way.

New institutions, initiatives, and collaborations are required – scientists sitting alongside representatives of different social and cultural groups, environmental campaigners letting go of the 'ownership' of climate change, and working through existing social networks. The purpose of this new infrastructure would be to catalyse new conversations about climate change. These would not be designed to win an argument, but to allow people to express and discuss their concerns, fears, dreams, and hopes for the future, providing answers to the ultimate question posed by climate change, 'how should we live?' (Corner and Groves 2014; Rapley et al. 2014).

This is a question that everyone has a stake in. It will never be conclusively answered: it is an ongoing discussion and negotiation. But we can do a much better job of providing opportunities for people to take part in these discussions. Despite being dwarfed by the financial muscle of industrial lobbyists, the combined budgets of NGOs, civil society groups and members' organisations with an interest in climate change at a global level are considerable (Dauvergne and Lebaron 2014). Governments pour many millions into physical infrastructure (including energy projects), but for the most part only pay lip service to public participation in climate policies.

From arranging town-hall meetings, to support and advice for existing social networks (online or in-person), to sponsoring debates and dialogue that explicitly reaches beyond the usual suspects and locates climate conversations somewhere entirely new, the logistical challenges would be considerable but by no means unachievable. And while there is certainly no time in a rapidly changing climate for dithering while temperatures rise, there is also no merit in advocating for climate policies that do not have broad-based social consent. An investment in the social infrastructure for climate conversations would produce a vibrant and dynamic public discourse on energy and climate change. And like all good investments, its value would become apparent over time, as individual campaigns, initiatives, and communication strategies found an increasingly attentive audience.

At the time of writing, the UN Paris agreement is being rapidly ratified by individual nations. There are increasingly positive signs that the global energy system is beginning to turn decisively away from fossil fuels and towards renewable technologies. These developments are welcome. But in other respects, there is little to celebrate, with extreme weather magnified by climate change on the rise, no recognition of the considerable impact of aviation and shipping in the Paris accord, and targets for the end of the century that imply either complete decarbonisation of existing systems and practices or the introduction of untested and unproven 'negative emissions' technologies. In key nations like the US and Australia, a poisonously polarised debate on climate change and energy choices persists.

But while it is possible to view the post-Paris landscape through an optimistic or pessimistic lens, it is unarguable that almost all of the heavy lifting on energy and climate change lies ahead of us. It is difficult to see how we can rise to this challenge without sustained and substantive public engagement. At present, though, there is a stifling social silence, with a muted majority sandwiched between narrow bands of activists at either end of the debate.

The evidence reviewed in this book offers tools, methods, and principles to break the social silence. New voices to catalyse engagement, new stories that resonate with diverse public values, and the cultivation of climate citizenship: these are the principles that can lift the energy and climate change discourse out of the margins and into the mainstream. It is time to start talking climate.

NOTES

1. Roberts, D. (2016) Is it worth trying to 'reframe' climate change? Probably not, Vox Energy and Environment [Online], Available from: http://www.vox.com/2016/3/15/11232024/reframe-climate-change. Accessed 23 June 2016.
2. http://www.climateactionreserve.org/blog/2012/08/31/environmental-cartoons-by-joel-pett/. Accessed 23 June 2016.
3. http://www.gov.scot/Topics/Environment/climatechange/scotlands-action/climatechangeact. Accessed 23 June 2016.
4. www.carbonconversations.org. Accessed 23 June 2016.
5. http://www.mumsnet.com/. Accessed 23 June 2016.
6. http://unfccc.int/cooperation_support/education_outreach/overview/items/8946.php. Accessed 23 June 2016.
7. http://my2050.decc.gov.uk/. Accessed 23 June 2016.

Glossary of Key Terms

Climate citizenship A sense of collective responsibility for meeting or exceeding national carbon targets, and a process of reflection on what they mean for people's lives, are the central components of climate citizenship.

Environmentalism A philosophy, ideology, or movement centred on protection of the natural environment. Although a concern for the natural environment has been a feature of systems of thought for millennia, our use of the term in this book is mostly with reference to the popular environmentalism that developed in Western culture in the 1960s and 1970s.

Frames The same information, when given a different label or title, or when linked to a particular theme or idea, can be perceived very differently. Frames bring certain elements of a complex issue to the fore, allowing people to make sense of it.

Narrative Narratives are stories and a key method by which people make sense of the world, learn values, form beliefs, and give shape to their lives.

Nudge A popular form of the discipline of behavioural economics, the nudge approach is based on making small changes to the environment in which choices are made, 'nudging' behaviours in the right direction as a consequence. An example of a nudge is changing the default choice for consumers, for example, being automatically 'opted-in' to a green energy tariff, rather than the default setting being 'opt out'.

© The Author(s) 2017
A. Corner, J. Clarke, *Talking Climate*,
DOI 10.1007/978-3-319-46744-3

Public engagement What people think, feel, and do about climate change. The ideas in this book are principles for widening and deepening public engagement.

Rebound effects A rebound effect describes the phenomenon where savings in energy use generated by one energy-saving behaviour are offset, reduced, or eliminated entirely through increased energy use elsewhere. A good example of a 'rebound' effect would be individuals deciding to 'treat themselves' to a foreign holiday with the money they had saved on their energy bill through insulation measures.

Sceptics/scepticism Climate change sceptics are people who dismiss or downplay the reality or seriousness of human-caused climate change.

Social marketing Social marketing involves the application of advertising and marketing techniques to achieve specific behavioural and social goals.

Spillover 'Spillover' is a word used to describe the idea that people who adopt one environmental behaviour are more likely to adopt other environmental behaviours. An example of spillover might involve someone applying behaviours practised in the workplace to their personal life.

Values A 'guiding principle in the life of a person', values are distinct from beliefs or attitudes, in that they are relatively stable constructs that people bring to bear on many different situations.

References

Abroms, L.C., & Maibach, E. (2008). The effectiveness of mass communication to change public behaviour. *Annual Review of Public Health*, 29, 219–234.

Alexander, J., Crompton, T., & Shrubsole, G. (2011). Think of me as evil? Opening the ethical debates in advertising, Public Interest Research Centre (PIRC) WWF-UK.

Austin, A., Cox, J., Barnett, J., & Thomas, C. (2011). Exploring catalyst behaviours: Full Report. A report to the Department for Environment, Food and Rural Affairs. Brook Lyndhurst for Defra, London.

Bain, P.G., Hornsey, M.J., Bongiorno, R., & Jeffries, C. (2012). Promoting pro-environmental action in climate change deniers. *Nature Climate Change*, 2, 600–603.

Bashir, N., Lockwood, P., Chasteen, A., Nadolny, D., & Noyes, I. (2013). The ironic impact of activists: Negative stereotypes reduce social change influence. *European Journal of Social Psychology*, 43(7), 614–626.

Bernauer, T., & McGrath, L. (2016). Simple reframing unlikely to boost public support for climate policy. *Nature Climate Change*. doi:10.1038/NCLIMATE2948.

Blackmore, E., Sanderson, B., & Hawkins, R. (2014). Valuing Equality: How equality bodies can use values to create a more equal & accepting Europe. Public Interest Research Centre.

Bolderdijk, J.W., Steg, L., Geller, E.S., Lehman, P.K., & Postmes, T. (2013). Comparing the effectiveness of monetary versus moral motives in environmental campaigning. *Nature Climate Change*, 3, 413–416.

BOND (2015). The Narrative Project: User Guide, BOND. Available at https://www.bond.org.uk/sites/default/files/resource-documents/the_narrative_project_user_guide_u.k._final_0.pdf. Accessed 30 May 2016.

© The Author(s) 2017
A. Corner, J. Clarke, *Talking Climate*,
DOI 10.1007/978-3-319-46744-3

Bourne, A. (2010) The role of fear in HIV prevention, Sigma Research. Available at http://sigmaresearch.org.uk/files/MiC-briefing-1-Fear.pdf. Accessed 30 May 2016.

Boykoff, M. (2011). *Who speaks for the climate? Making sense of media reporting on climate change.* Cambridge: Cambridge University Press.

Brand, C., & Boardman, B. (2008). Taming of the few – the unequal distribution of greenhouse gas emissions from personal travel in the UK. *Energy Policy*, 36, 224–238.

Brookman, David E., & Joshua L. Kalla. (2016). Durably reducing transphobia: A field experiment on door-to-door canvassing. *Science*, 352(6282), 220–224.

Brügger, A., Dessai, S., Devine-Wright, P., Morton, T.A., & Pidgeon, N.F. (2015) Psychological responses to the proximity of climate change. *Nature Climate Change*, 5, 1031–1037.

Brulle, R.J. (2010). From environmental campaigns to advancing the public dialog: Environmental communication to civic engagement. *Environmental Communication*, 4(1), 82–98.

Brulle, R.J., Carmichael, J., Jenkins, C. (2012). Shifting public opinion on climate change: An empirical assessment of factors influencing concern over climate change in the U.S., 2002–2010. *Climatic Change*, 114, 169–188.

Bushell, S., Workman, M., & Colley, T. (2016) Towards a unifying narrative for climate change, Briefing Paper No. 18, Grantham Institute, Imperial College London. Available at https://www.imperial.ac.uk/media/imperial-college/grantham-institute/public/publications/briefing-papers/Towards-a-unifying-narrative-for-climate-change-Grantham-BP18.pdf. Accessed 30 May 2016.

Campbell, T., & Kay, A. (2014). Solution aversion: On the relation between ideology and motivated disbelief. *Journal of Personality and Social Psychology*, 107(5), 809–824.

Capstick, S., & Lewis, A. (2008). *Personal carbon trading: Perspectives from psychology & behavioural economics.* London: IPPR.

Capstick, S.B., Demski, C.C., Sposato, R.G., Pidgeon, N.F., Spence, A., & Corner, A. (2015a). Public perceptions of climate change in Britain following the winter 2013/2014 flooding. Understanding Risk Research Group Working Paper 15-01, Cardiff University, Cardiff, UK.

Capstick, S., Lorenzoni, I., Corner, A., & Whitmarsh, L. (2015b). Prospects for radical emissions reduction through behavior and lifestyle change. *Carbon Management*, 5(4), 429-445.

Carbon Brief (2013). Opinion research results, Available at http://www.carbonbrief.org/blog/2013/08/two-degrees-don%E2%80%99t-you-mean-eightol ling-shows-people-think-dangerous-climate-changemeans-eight-degrees-of-warming/. Accessed Dec 18 2014.

Center for Research on Environmental Decisions and ecoAmerica. (2014). *Connecting on climate: A guide to effective climate change communication*, 42. New York/Washington, DC Available at http://ecoamerica.org/wp-content/uploads/2014/12/ecoAmerica-CRED-2014-Connecting-on-Climate.pdf.

Chitnis, M., Sorrell, S., Druckman, A., Firth, S.K., & Jackson, R. (2013). Turning light into flights: Estimating direct and indirect rebounds effect for UK households. *Energy Policy*, 55, 234–250.

Christakis, N., & Fowler, J. (2009). *Connected*. New York: Little, Brown and Co.

Church of England (2004) Christian agencies unite to make poverty history in 2005 [press release] 22 December 2004, Available at https://www.churchofengland.org/media-centre/news/2004/12/makepoverthistory.aspx. Accessed 30 May 2016.

Cialdini, R.B. (2003). Crafting messages to protect the environment. *Psychological Science*, 12(4), 105–109.

Common Cause Foundation. (2016). *Perceptions matter: The common cause UK values survey*. London: Common Cause Foundation.

Cook, J., Nuccitelli, D., Green, S. A., Richardson, M., Winkler, B., Painting, R., Way, R., Jacobs, P., & Skuce, A. (2013). Quantifying the consensus on anthropogenic global warming in the scientific literature. *Environmental Research Letters*, 8, 2.

Corner, A., Pidgeon, N., & Parkhill, K. (2012). Perceptions of geoengineering: Public attitudes, stakeholder perspectives, and the challenge of 'upstream' engagement. *WIREs Climate Change*, 3(5), 451–466.

Corner, A. (2013a). *A new conversation with the centre-right about climate change: Values, frames and narratives*. Oxford: Climate Outreach & Information Network.

Corner, A. (2013b). *Climate silence (and how to break it)*. Oxford: Climate Outreach & Information Network.

Corner, A., Lewandowsky, S., Phillips, M., & Roberts, O. (2015a). *The uncertainty handbook*. Bristol: University of Bristol.

Corner, A., Roberts, O., Chiari, S., Völler, S., Mayrhuber, E., Mandl, S., & Monson, K. (2015b). How do young people engage with climate change? The role of knowledge, values, message framing, and trusted communicators. *WIREs Climate Change*, 6(5), 523–534.

Corner, A., Webster, R. & Teriete, C. (2015c). *Climate Visuals: Seven principles for visual climate change communication (based on international social research)*. Oxford: Climate Outreach.

Corner, A., Marshall, G., & Clarke, J. (2016). *Communicating effectively with the centre-right about household energy-efficiency and renewable energy technologies*. Oxford: Climate Outreach.

Corner, A., & Groves, C. (2014). Breaking the climate change communication deadlock. *Nature Climate Change*, 4(9), 743–745.

Corner, A., & van Eck, C. (2014). Science and stories: Bringing the IPCC to life. *Climate Outreach*.

Corner, A. J., Markowitz, E., & Pidgeon, N. F. (2014). Public engagement with climate change: The role of human values. *Wiley Interdisciplinary Reviews: Climate Change*, 5(3), 411–422.

Corner, A.J. & Randall, A. (2011). Selling climate change? The limitations of social marketing as a strategy for climate change public engagement. *Global Environmental Change*, 21(3), 1005–1014.

Corner, A.J., & Roberts, O. (2014). *How narrative workshops informed a national climate change campaign*. Oxford: Climate Outreach & Information Network.

Crompton, T. (2010). *Common cause: The case for working with our cultural values*. UK: WWF.

Crompton, T., Kasser, T. (2009). *Meeting environmental challenges: The role of human identity*. Surrey: WWF UK.

Darnton, A., & Kirk, M. (2011) Finding Frames: New Ways to Engage the UK Public in Global Poverty, BOND. Available at http://www.findingframes.org/Finding%20Frames%20New%20ways%20to%20engage%20the%20UK%20public%20in%20global%20poverty%20Bond%202011.pdf. Accessed 30 May 2016.

Das, E.H., de Wit, J.B., & Stroebe, W. (2003). Fear appeals motivate acceptance of action recommendations: Evidence for a positive bias in the processing of messages. *Personality and Social Psychology Bulletin*, 29(5), 650–664.

Dauvergne, P., & Lebaron, G. (2014). *Protest Inc: The corporatization of activism*. Cambridge: Polity.

De Groot, J., & Steg, L. (2008). Value orientations to explain beliefs related to environmental significant behavior how to measure egoistic. *Altruistic, and Biospheric Value Orientations, Environment and Behaviour*, 40(3), 330–354.

Devine-Wright, P. (2007). Reconsidering public attitudes and public acceptance of renewable energy technologies: A critical review, published by the School of Environment and Development, University of Manchester. Available at http://www.sed.manchester.ac.uk/research/beyond_nimbyism/. Accessed 23 June 2016.

Dietz, T., Gardner, G. T., Gilligan, J., Stern, P. C., Vandenbergh, M. P. (2009a). Household actions can provide a behavioral wedge to rapidly reduce US carbon emissions. *Proceedings of National Academy Sciences*, 106(44), 18452–18456.

Dietz, T., Stern, P.C., & Dan, A. (2009b). How deliberation affects stated willingness to pay for mitigation of carbon dioxide emissions: An experiment. *Land Economy*, 85, 329–347.

Dobson, A. (2003). *Citizenship and the environment*. Oxford: Oxford University Press.

Dobson, A., (2010). *Environmental Citizenship and Pro-environmental Behaviour*. Rapid Research and Evidence Review. Sustainable Development Research Review: London.

Doherty, B. (2003). 'The Preferred Way of Doing Things': The British Direct Action Movement. *Parliamentary Affairs*, 56(4), 669–686.

Douglas, M., & Wildavsky, A. B. (1982). *Risk and culture: An essay on the selection of technical and environmental dangers.* California: University of California Press.

Downs, J., de Bruin, W., & Fischhoff, B. (2008). Parents' vaccination comprehension and decisions. *Vaccine*, 26(12), 1595–1607.

Doyle, J. (2011). *Mediating climate change.* Ashgate: Farnham.

Ebeling, F; Lotz, S. (2015). Domestic uptake of green energy promoted by opt-out tariffs. *Nature Climate Change*, 5, 868–871.

ECIU – Energy & Climate Intelligence Unit. (2014). Study shows widespread misconceptions about energy and climate change. Retrieved from http://eciu.net/press-releases/2014/survey-reveals-widespread-misconceptions-about-energy-and-climate-change. Accessed 23 June 2016.

European Centre for Disease Prevention and Control. (2012). *Communication on immunisation – Building trust.* Stockholm: ECDC.

Evans, L., Gregory, R.M., Corner, A., Hodgetts, J., Ahmed S., & Hahn, U. (2013). Self Interest and pro-environmental behaviour. *Nature Climate Change*, 3, 122–125.

Feinberg, M., & Willer, R. (2013). The moral roots of environmental attitudes. *Psychological Science*, 24, 56–62.

Fielding, K.S; Hornsey, M.J. (2016). A cautionary note about messages of hope: Focusing on progress in reducing carbon emissions weakens mitigation motivation. *Global Environmental Change*, 39, 26–34.

Fischer, E. M., & Knutti, R. (2015). Anthropogenic contribution to global occurrence of heavy-precipitation and high-temperature extremes. *Nature Climate Change*, 5, 560–565.

Geiger, N., & Swim, J. (2016). Climate of silence: Pluralistic ignorance as a barrier to climate change discussion. *Journal of Environmental Psychology*, 47, 79–90.

Gifford, R., & Comeau, L.A. (2011). Message framing influences perceived climate change competence, engagement, and behavioural intentions. *Global Environmental Change*, 21, 1301–1307.

Gill Ereaut, G., & Segnit, N. (2006). *Warm Words How are we telling the climate story and can we tell it better?* IPPR.

Grundmann, R. (2016). Climate change as a wicked social problem. *Nature Geoscience*, 9, 562–563.

Hagendijk, R., & Irwin, A. (2006). Public deliberation and governance: engaging with science and technology in contemporary Europe. *Minerva*, 44(2), 167–184.

Haidt, J. (2007). The new synthesis in moral psychology. *Science*, 316, 998–1002.

Hastings, G. (2007). *Social marketing: Why should the devil have all the best tunes?* Oxford: Elsevier.

Henson, S., Lindstrom, J., Haddad, L., & Mulmi, R. (2010). Public perceptions of international development and support for aid in the UK: Results of a qualitative enquiry. *IDS Working Papers*, 2010(353),pp. 01–67.

Henwood, K., Pidgeon, N., Groves, C., Shirani, F., Butler, C., & Parkhill, K. (2015) Energy Biographies Research Report. Available at http://energybiographies. org/our-work/our-findings/reports/. Accessed 19 May 2016.

Herman, D. (2013). *Narrative theory and the cognitive sciences.* Chicago: Center for the Study of Language and Information.

Hobson, K., & Niemeyer, S. (2012). "What sceptics believe": The effects of information and deliberation on climate change scepticism. *Public Understanding of Science, 22*(4), 396–412.

Hogg, M., & Shah, H. (2010). *The impact of global learning on public attitudes and behaviours towards international development and sustainability.* London: Development Education Association.

Holmes, T., Blackmore, E., Hawkins, R., & Wakeford, T. (2011). *The common cause handbook.* Public Interest Research Centre.

Hoog, N., Stroebe, W., & de Wit, J.B.F. (2005). The impact of fear appeals on processing and acceptance of action recommendations. *Personality & Social Psychology Bulletin, 31,* 24–33.

Hoppner, C., Whitmarsh, L. (2010). Public and policy expectations regarding public engagement in climate change action. In Whitmarsh, L., O'Neill, S., & Lorenzoni, I. (Eds.), *Engaging the public with climate change: Behaviour change and communication.* London: Earthscan.

Hornsey, M. J., Harris, E. A., Bain, P. G., & Fielding, K. S. (2016). Meta-analyses of the determinants and outcomes of belief in climate change. *Nature Climate Change, 6,* 622–626.

Hornsey, M. J., & Fielding, K. S. (2016). A cautionary note about messages of hope: Focusing on progress in reducing carbon emissions weakens mitigation motivation. *Global Environmental Change, 39,* 26–34.

Howell, R. A. (2014). Promoting lower-carbon lifestyles: The role of personal values, climate change communications and carbon allowances in processes of change. *Environmental Education Research, 20*(3), 434–435.

Hulme, M. (2009). *Why we disagree about climate change: Understanding controversy, inaction and opportunity.* Cambridge, UK: Cambridge University Press.

Inglehart, R.F. (2008). Changing values among western publics from 1970 to 2006. *Western European Politics, 31,* 130–146.

Involve (2010) Nudge, think or shove? Shifting values and attitudes towards sustainability: A briefing for sustainable development practitioners, November 2010, Available at. http://www.involve.org.uk/wp-content/uploads/2011/03/Nudge-think-or-shove.pdf. Accessed 23 June 2016.

Isaksson, T., & Corner, A. (2016) Webinar and guide: Managing the psychological distance of climate change, Climate Outreach, Available at http://climateoutreach.org/resources/psychological-distance/. Accessed 27 May 2016.

Jones, M., & Song, G. (2014). Making sense of climate change: How story frames shape cognition. *Political Psychology*, 35(4), 447–476.

Kahan D. (2012). Why we are poles apart on climate change? *Nature*, 488, 255.

Kahan, D., Peters, E., Wittlin, M., Slovic, P., Ouellette, L. L., Braman, D., & Mandel, G. (2012). The polarizing impact of science literacy and numeracy on perceived climate change risks. *Nature Climate Change*, 2, 732–735.

Kasser T., & Crompton T. (2009). *Meeting environmental challenges: The role of human identity*. WWF UK.

Kelley, C. P., Mohtadi, S., Cane, M. A., Seager, R., & Kushnir, Y. (2015). Climate change in the fertile crescent and implications of the recent Syrian drought. *PNAS*, 11, 3241–3246.

Kurz, T., Gardner, B., Verplanken, B., & Abraham, C. (2014). Habitual behaviors or patterns of practice? Explaining and changing repetitive climate-relevant actions. *WIREs Clim Change*, 6(1), 113–128.

Lakoff, G. (1990). *Don't think of an elephant: Know your values and frame the debate*. Vermont: Chelsea Green Publishing.

Land Use Consultants (2011). Climate change conversations. Scottish Natural Heritage Commissioned Report No. 492.

Lazer, W., Kelley, E.J. Eds. (1973). *Social marketing: Perspectives and viewpoints*. Ontario: Irwin-Dorsey.

Leiserowitz, A. (2006). Climate change risk perception and policy preferences: The role of affect, imagery, and values. *Climatic Change*, 77, 45–72.

Leiserowitz, A., Maibach, E., Roser-Renouf, C. & Smith, N. (2011). Global Warming's Six Americas. Yale University and George Mason University. New haven, CT: Yale Project on Climate Change Communication. Retrieved from http://environment.yale.edu/climatecommunication/files/SixAmericas May2011.pdf. Accessed 23 June 2016.

Leiserowitz, A., Feinberg, G., Rosenthal, S., Smith, N., Anderson A., Roser-Renouf, C., & Maibach, E. (2014). *What's in a name? Global warming vs. Climate change*. Yale Project on Climate Change Communication New Haven: Yale University and George Mason University.

Leiserowitz, A., Maibach, E., Roser-Renouf, C., Feinberg, G., & Rosenthal, S. (2015). *Climate change in the American mind: March, 2015*. Yale University and George Mason University. New Haven: Yale Project on Climate Change Communication. Retrieved from http://climatecommunication.yale.edu/wp-content/uploads/2015/04/Global-Warming-CCAM-March-2015.pdf. Accessed 23 June 2016.

Leombruni, L. (2015). How you talk about climate change matters: A communication network perspective on epistemic skepticism and belief strength. *Global Environmental Change*, 35, 148–161.

Leviston, Z., Walker, I., & Morwinski, S. (2012). Your opinion on climate change might not be as common as you think. *Nature Climate Change*, 3, 334–337.

Lewandowsky, S., Gignac, G. & Vaughan, S. (2012). The pivotal role of perceived scientific consensus in acceptance of science. *Nature Climate Change*, 3, 399–404.

Lewandowsky, S., Gilles, G., & Vaughan, S. (2013). The pivotal role of perceived scientific consensus in acceptance of science. *Nature Climate Change*, 3, 399–404.

Lockwood, M. (2013). The political sustainability of climate policy: The case of the UK Climate Change Act. *Global Environmental Change*, 23(5), 1339–1348.

Lorenzoni, I., Nicholson-Cole, S., & Whitmarsh, L. (2007). Barriers perceived to engaging with climate change among the UK public and their policy implications. *Global Environment Change*, 17(3–4), 444–459.

Lorenzoni, I., & Pidgeon, N.F. (2006). Public views on climate change: European and USA perspectives. *Climate Change*, 77, 73–95.

Luber, G., & Lemery, J. (2015). *Global climate change and human health: From science to practice*. San Francisco: Wiley.

Machin, A. (2013). *Negotiating climate change: Radical democracy and the illusion of consensus*. London: Zed books.

Mahoney, J. (2010). Strategic communication and anti-smoking campaigns. *Public Communication Review*, 1, 33–48.

Maibach, E., Myers, T., &. Leiserowitz, A. (2014). Climate scientists need to set the record straight: There is a scientific consensus that human-caused climate change is happening. *Earth's Future*, 2(5), 295–298.

Maibach, E.W., Nisbet, M., Baldwin, P., Akerlof, K., & Diao, G. (2010). Reframing climate change as a public health issue: An exploratory study of public reactions. *BMS Public Health*, 10(1), 299–309.

Maio, G.R. (2011). Don't mind the gap between values and action. Discussion Paper. Common Cause. Retrieved at http://valuesandframes.org/down load/briefings/Value-Action%20Gap%20|%20Common%20Cause% 20Briefing.pdf. Accessed 23 June 2016.

Maio, G. R. (2015). The psychology of human values. In *European monographs in social psychology*. London: Psychology Press.

Manzo, K. (2010). Beyond polar bears? Re-envisioning climate change. *Meteorological Applications*, 17, 196–208.

Markowitz, E.M., & Shariff, A.F. (2012). Climate change and moral judgement. *Nature Climate Change*, 2, 243–247.

Marshall, G. (2014a). *Don't even think about it: Why our brains are wired to ignore climate change*. New York: Bloomsbury.

Marshall, G. (2014b). *After the floods: Communicating climate change around extreme weather*. Oxford: Climate Outreach Information Network.

Marshall, G. (2014c). *Hearth and Hiraeth: Constructing climate change narratives around national identity*. Oxford: Climate Outreach Information Network.

Marshall, G. (2014d). You say "Global Warming". I say "Climate Change"- Let's call the whole thing off! Talking Climate blog post, 29 May. Retrieved from http://talkingclimate.org/you-say-global-warming/.

Marshall, G. (2015). *Starting a new conversation on climate change with the European centre-right*. Oxford: Climate Outreach Information Network.

Marshall, G., Corner, A., & Clarke, J. (2016). *Positive visions of climate change*. 4 Stories: City Home: Countryside and Locality for a mainstream centre-right audience. Climate Outreach.

Marshall, G., & Darnton, A. (2012). Sustainable Development Narratives for - Wales: A Toolkit for Government Communications. Welsh Government.

McCright, A. et al. (2015). Examining the effectiveness of climate change frames in the face of a climate change denial counter-frame. *Topics in Cognitive Science*, 8(1), 76–97.

McCright, A.M., & Dunlap, R.E. (2011). Cool dudes: The denial of climate change among conservative white males in the United States. *Global Environmental Change*, 21(4), 1163–1172.

McDonald, R., Chai, H.Y., & Newell, B.R. (2015). Personal experience and the 'psychological distance' of climate change: An integrative review. *Journal of Environmental Psychology*, 44(1), 9–118.

McNeill, K., Doane, D., & Tarman, G. (2012) The lessons from Make Poverty History, The Guardian 31 May 2012, Available at http://www.theguardian.com/voluntary-sector-network/2012/may/31/make-poverty-history. Accessed 30 May 2016.

Menz, F. C., & Seip, H. M. (2004). Acid rain in Europe and the United States. *An Update, Environmental Science and Policy*, 7, 253–265.

Moser, S (2016). Reflections on climate change communication research and practice in the second decade of the 21st century: What more is there to say? WIREs – Climate Change, Available at http://onlinelibrary.wiley.com/doi/10.1002/wcc.403/abstract. Accessed 22 April 2016.

Moser S.C., & Dilling L. (2007). *Creating a climate for change: Communicating climate change and facilitating social change*. Cambridge, UK: Cambridge University Press.

Myers, T.A., Nisbet, M.C., Maibach, E.W., & Leiserowitz, A.A. (2012). A public health frame arouses hopeful emotions about climate change. *Climatic Change*, 113(3–4), 1105–1112.

Nagelkerken, I., & Connell, S. D. (2015) Global alteration of ocean ecosystem functioning due to increasing human CO_2 emissions, Proceedings of the Natural Academy of Sciences. USA, 112, 13272–13277.

Nash, N. et al. (2012). Developing Narratives for a Sustainable Wales: Focus Group Synthesis Report. Welsh Government.

Nerlich, B., Koteyko, N., & Brown, B. (2010). Theory and language of climate change communication. *WIREs Climate Change*, 1, 97–110.

Nisbet, M.C. (2009). Communicating climate change: Why frames matter for public engagement. *Environment*, 51(2), 12–25.

Norgaard, K. M. (2011). *Living in denial: Climate change, emotions and everyday life*. Massachusetts: MIT Press.

Norton, S. (2014) Rambling revolution: How people power won the right to roam, *The Independent* [online], 3 October 2016, Available at http://www.independent.co.uk/travel/uk/rambling-revolution-how-people-power-won-the-right-to-roam-9773740.html. Accessed 30 May 2016.

Nulman, E. (2015). *Climate change and social movements: Civil Society and the Development of National Climate Change Policy*. Palgrave Macmillan.

Nurmis, J. (2016). Visual climate change art 2005–2015: Discourse and practice. *WIREs Climate Change*, 7, 501–516.

Nye, M., & Burgess, J. (2008). *Promoting durable change in household waste and energy use behaviour*. London: Department for Environment, Food & Rural Affairs.

O'Neill, S., & Nicholson-Cole, S. (2009). "Fear Won't Do It" Promoting positive engagement with climate change through visual and iconic representations. *Science Communication*, 30(3), 355–379.

O'Neill, S.J., & Hulme, M. (2009). An iconic approach for representing climate change. *Global Environmental Change*, 1(9), 402–410.

Ockwell, D., Whitmarsh, L., & O'Neil, S. (2009). Reorienting climate change communication for effective mitigation: forcing people to be green or fostering grassroots engagement? *Science Communication*, 30, 305–327.

Oreskes, N., & Conway, E. M. (2012). *Merchants of doubt: How a handful of scientists obscured the truth on issues from tobacco smoke to global warming*. New York: Bloomsbury Press.

P. John, & G. Stoker, (2010) How experiments can help get Britain to the Big Society. Available at http://www.civicbehaviour.org.uk/documents/findin goffindingsformatted_002.pdf. Accessed 30 May 2016.

Painter, J. (2011). *Poles apart – The international reporting of climate scepticism*. Oxford: Reuters Institute for the Study of Journalism, University of Oxford.

Painter, J. (2015). Taking a bet on risk. *Nature Climate Change*, 15(4), 288–289.

Painter, J., & Gavin, N.T. (2015). Climate Skepticism in British Newspapers, 2007–2011. *Environmental Communication*, 10, 432–452.

Panos London. (2003). *Missing the message? 20 years of learning from HIV/AIDS*. London: The Panos Institute.

Parkhill, K.A., Demski, C., Butler, C., Spence, A., & Pidgeon, N. (2013). *Transforming the UK energy system: Public values, attitudes and acceptability: synthesis report*. London: UKERC.

Peattie, K., & Peattie, S. (2009). Social marketing: A pathway to consumption reduction? *Journal of Business Research*, 62(2), 260–268.

Petts, J., & Niemeyer, S. (2004). Health risk communication and amplification: learning from the MMR vaccination controversy. *Health, Risk & Society*, 6(1), 7–23.

Petty, R. E., & Cacioppo, J. T. (1984). Source factors and the elaboration likelihood model of persuasion. *Advances in Consumer Research*, 11, 668–672.

Pidgeon, N., Demski, C., Butler, C., Parkhill, K., & Spence, A. (2014). Creating a national citizen engagement process for energy policy. *Proceedings of the National Academy of Sciences*, 111(Supplement_4), 13606–13613.

Pidgeon, N.F. (2012). Public understanding of, and attitudes to, climate change: UK and international perspectives and policy. *Climate Policy*, 12(S1), S85–S106.

Pidgeon, N.F., & Fischhoff, B. (2011). The role of social and decision sciences in communicating uncertain climate risks. *Nature Climate Change*, 1, 35–41.

Rabinovich, A., Morton, T.A., & Duke, C.C. (2010). Collective self and individual choice: The role of social comparisons in promoting climate change. In Whitmarsh, L., O'Neill, S. & Lorenzoni, I. (Eds.), *Engaging the public with climate change: Behaviour change and communication*. London: Earthscan.

Rabinovich, A., Morton, T.A., & Birney, M.E. (2012). Communicating climate science: The role of perceived communicator's motives. *Journal of Environmental Psychology*, 32, 11–18.

Rapley, C. G., de Meyer, K., Carney, J., Clarke, R., Howarth, C., Smith, N., Stilgoe, J., Youngs, S., Brierley, C., Haugvaldstad, A., Lotto, B., Michie, S., Shipworth, M. &, Tuckett, D. (2014) Time for Change? Climate Science Reconsidered, Report of the UCL Policy Commission on Communicating Climate Science.

Renn, O., Webler, T., & Wiedemann, P. (1995). The pursuit of fair and competent citizen Participation. In *Fairness and competence in citizen participation* (pp. 339–367). the Netherlands: Springer.

Reser, J.P., Bradley, G.L., & Ellul, M.C. (2014). Encountering climate change: 'Seeing' is more than believing'. *Wiley Interdisciplinary Reviews: Climate Change*, 5(4), 521–537.

Revkin, A. (2008). Global heating, Atmosphere Cancer, Pollution Death. What's in a Name? *The New York Times* 18 February. Retrieved from: http://dotearth.blogs.nytimes.com/2008/02/18/global-heating-atmosphere-cancer-pollution-death-whats-in-a-name/?_php=true%26_type=blogs%26_r=0#h. Accessed 23 June 2016.

Rhodes, E., Axsen, J., Jaccard, M. (2014). Does effective climate policy require well-informed citizen support? *Global Environmental Change*, 29, 92–104.

Rowe, G., & Frewer, L.J. (2005). A typology of public engagement mechanisms. *Science, Technology & Human Values*, 30(2), 251–290.

Rowson, J., Broome, S., Jones, A. (2010). *Connected communities: How social networks power and sustain the big society*. London: Royal Society of Arts.

Rowson, J. (2013). *A new agenda on climate change: Facing up to stealth denial and winding down on fossil fuels.* London: Royal Society of Arts.

Rowson, J. (2015) Money talks, divest invest and the battle for climate realism, RSA Available at https://www.thersa.org/discover/publications-and-articles/reports/money-talks—divest-invest-and-the-battle-for-climate-realism. Accessed 30 May 2016.

Rowson, J., & Corner, A. (2015) The seven dimensions of climate change: Introducing a new way to think, talk and act, Available at http://climateoutreach.org/resources/the-seven-dimensions-of-climate-change-introducing-a-new-way-to-talk-think-and-act/. Accessed 19 May 2016.

Sachs, J. (2005). *The end of poverty.* New York: Penguin Press.

Schlembach, R. (2011). How do radical climate movements negotiate their environmental and their social agendas? A study of debates within the Camp for Climate Action (UK). *Critical Social Policy,* 31(2), 194–215.

Schuldt, J.P., Konrath, S.H., & Schwartz, N. (2011). "Global Warming" or "Climate Change"? Whether the planet is warming depends on question wording. *Public Opinion Quarterly,* 75(1), 115–124.

Schwartz, S.H. (1992). Universals in the content and structure of values: Theoretical advances and empirical tests in 20 countries. Cambridge, MA: Academic Press, Inc.

Schwartz, S.H., Cieciuch, J., Vecchione, M., Davidov, E., Fischer, R., Beierlein, C., Ramos, A., Verkasalo, M., Lönnqvist, J.E., Demirutku, K., Dirilen-Gumus, O., Konty, M. (2012). Refining the theory of basic individual values. *Journal of Personality and Social Psychology,* 103, 663–688.

Scruggs, L., & Benegal, S. (2012). Declining public concern about climate change: Can we blame the great recession? *Global Environmental Change,* 22(2), 505–515.

Shamir, J., & Shamir, M. (1997). Pluralistic ignorance across issues and over time: Information cues and biases. *Public Opinion Quarterly,* 61, 227–260.

Shaw, C. (2015). *The two degrees dangerous limit for climate change: Public understanding and decision making.* Abingdon: Routledge.

Shaw, C. Hellsten, I, & Nerlich, B. (2016). Framing risk and uncertainty in social science articles on climate change. In 1995–2012 in communicating risk, Eds. Crichton, J. Candlin, C. and Firkins A.S. Palgrave Macmillan.

Shaw, C., & Corner, A. (2016). *Climate change conversation series: Desk review.* April 2016 Oxford: Climate Outreach.

Sheldon, K.M, & Nichols, C.P. (2009). Comparing Democrats and Republicans on intrinsic and extrinsic values. *Journal of Applied Social Psychology,* 39, 589–623.

Smith, J., Tyszczuk, R., & Butler, R. Eds. (2014). *Culture and climate change: Narratives. culture and climate change,* Vol. 2. Cambridge, UK: Shed.

Spence, A., Poortinga, W., & Pidgeon, N. (2012). The psychological distance of climate change. *Risk Analysis,* 32(6), 957–972.

Spence, A., & Pidgeon, N. F. (2010). Framing and communicating climate change: The effects of distance and outcome frame manipulations. *Global Environmental Change*, 20(4), 656–667.

Spratt, D. (2012). Bright-siding climate advocacy and its consequences. *Climate Code Red*.

Spurling N, McMeekin A, Shove E, Southerton D, & Welch D. (2013). Interventions in practice: re-framing policy approaches to consumer behaviour. Sustainable Practices Research Group.

Stavins, R., Chan, G., Stowe, R., & Sweeney, R. (2012). The US sulphur dioxide cap and trade programme and lessons for climate policy. *VOX* [Online], Available at http://voxeu.org/article/lessons-climate-policy-us-sulphur-dioxide-cap-and-trade-programme. Accessed 30 May 2016.

Stokes, B., Wike, R., & Carle, J. (2015) Global Concern about Climate Change, Broad Support for Limiting Emissions, U.S., China Less Worried; Partisan Divides in Key Countries, Pew Research Center, [online] Nov 5. Available at http://www.pewglobal.org/2015/11/05/global-concern-about-climate-change-broad-support-for-limiting-emissions/. Accessed 25 May 2016.

Stoknes, P. E. (2015). *What we think about when we try not to think about global warming: Toward a new psychology of climate action*. USA: Chelsea Green Publishing.

Stoll-Kleemann, S., O'Riordan, T., & Jaeger, C. (2001). The psychology of denial concerning climate mitigation measures: Evidence from Swiss focus groups. *Global Environmental Change*, 11(2), 107–117.

Sturgis, P., & Allum, N. (2004). Science in society: Re-evaluating the deficit model of public attitudes. *Public Understanding of Science*, 13(1), 55–74.

Sunstein, C. R. and Thaler, R. H. (2009). *Nudge: Improving decisions about health, wealth and happiness*. London: Penguin.

Tarrow, S. (2011). *Power in movement*. Cambridge: Cambridge University Press.

Thaler, R. (2015). *Misbehaving: The making of behavioural economics*. Allen Lane.

Thøgersen, J., & Crompton, T. (2009). Simple and painless? The limitations of spillover in environmental campaigning. *Journal of Consumer Policy*, 32(2), 141-163.

Thøgersen, J. & Noblet, C.L. (2012). Does Green Consumerism Increase the Acceptance of Wind Power? *Energy Policy*, 51, 854-862.

Thomas, G.O., Poortinga, W. & Sautkina, E. (2016). The Welsh Single-Use Carrier Bag Charge and behavioural spillover. *Journal of Environmental Psychology*, 47, 126-135.

Tiefenbeck, V., Staake, T., Roth, K., & Sachs, O. (2013). For better or for worse? Empirical evidence of moral licensing in a behavioral energy conservation campaign. *Energy Policy*, 57, 160-171.

Tindall, D., & Piggot, G. (2015). Influence of social ties to environmentalists on public climate change perceptions. *Nature Climate Change*, 5(6), 546-549.

Topos Partnership (2009). Climate crossroads: A research-based framing guide. Topos Partnership.

van der Linden, S. (2014). On the relationship between personal experience, affect and risk perception: The case of climate change. *European Journal of Social Psychology*, 44(5), 430–440.

van der Linden, S. L., Leiserowitz, A. A., Feinberg, G. D., & Maibach, E. W. (2014). How to communicate the scientific consensus on climate change: Plain facts, pie charts or metaphors? *Climate Change*, 126, 255–262.

Verplanken, B., Walker, I., Davis, A., & Jurasek, M. (2008). Context change and travel mode choice: Combining the habit discontinuity and self-activation hypotheses. *Journal of Environmental Psychology*, 28, 121–127.

Villar, A., & Krosnick, J.A. (2011). Global warming vs. climate change, taxes vs. prices: Does word choice matter? *Climatic Change*, 105(1–2), 1–12.

Vulturius, G., Davis, M. & Bharwani, S. (2016). Building bridges and changing minds: Insights from climate communication research and practice. Stockholm Environmental Institute.

Wall, D. (2010). *The rise of the green left*. London: Pluto Press.

Weber, E. (2006). Experience-based and description-based perceptions of long-term risk: Why global warming does not scare us (Yet). *Climatic Change*, 77(1–2), 103–120.

Weber, E. (2010). What shapes perceptions of climate change? *Wiley Interdisciplinary Reviews: Climate Change*, 1(3), 332–342.

Weintrobe, S. (2012). *Engaging with climate change: Psychoanalytic and interdisciplinary perspectives*. Sussex: Routledge.

Weisbart, E. (2015) Pharmaceutical manufacturers create distrust of vaccinations, Physicians for a National Health Program [online] Available at http://www.pnhp.org/news/2015/february/pharmaceutical-manufacturers-create-distrust-of-vaccinations. Accessed 30 May 2016.

Weiss, M. (2012). *Is Acid Rain a Thing of the Past?* [online] Science |AAAS. Available at http://www.sciencemag.org/news/2012/06/acid-rain-thing-past. Accessed 30 May 2016.

Western Strategies & Lake Research Partners. (2009). *Climate and energy truths: Our common future*. Washington, US: EcoAmerica.

Whitmarsh, L. (2009). What's in a name? Commonalities and differences in public understanding of "climate change" and "global warming.". *Public Understanding of Science*, 18(4), 401–420.

Whitmarsh, L. (2011). Scepticism and uncertainty about climate change: Dimensions, determinants and change over time. *Global Environmental Change*, 21(2), 690–700.

Whitmarsh, L. E., O'Neill, S., Lorenzoni, I. (2011). *Engaging the public with climate change*. Abingdon: Earthscan.

Whitmarsh, L.E. & Corner, A. (forthcoming). Tools for a new climate conversation: A mixed-methods study of language for public engagement across the political spectrum.

Whitmarsh, L. E., & O'Neill, S. (2010). Green identity, green living? The role of pro-environmental self-identity in determining consistency across diverse pro-environmental behaviours. *Journal of Environmental Psychology*, 30(3), 305–314.

Wiebe, G.D. (1952). Merchandising commodities and citizenship on television. Public Opinion Quarterly 15, 679–691. Wiebe, G.D., 1952. Merchandising commodities and citizenship on television. *Public Opinion Quarterly*, 15, 679–691.

Witte, K., & Allen, M. (2000). A meta-analysis of fear appeals: implications for effective public health campaigns. Health, Education & Behaviour, 27(5), 591-615.

Wolf, J., Brown, K., Conway, D. (2009). Ecological citizenship and climate change: Perceptions and practice. *Environmental Politics*, 18(4), 503–521.

Wolsko, C., Ariceaga, H. & Seiden, J. (2016). Red, white, and blue enough to be green: Effects of moral framing on climate change attitudes and conservation behaviors. *Journal of Experimental Social Psychology*, 65, 7–19.

Wright, C. (2016). The best thing a business could do for the environment is shut down. *The Guardian*, Available at http://www.theguardian.com/sustainable-business/2016/jan/28/climate-change-capitalism-business-emotions-hea throw-protest-short-term-profits. Accessed 7 April 2016.

Wynne, B., & Irwin, A. eds. (1996). *Misunderstanding science?: The public reconstruction of science and technology*. Cambridge: Cambridge University Press.

Yaqub, O., Castle-Clarke, S., Sevdalis, N., & Chataway, J. (2014). Attitudes to vaccination: A critical review. *Social Science & Medicine*, 112, 1–11.

Yusoff, K., & Gabrys, J. (2011). Climate change and the imagination. *WIREs Climate Change*, 2, 516–534.

Zia, A., and Todd, A.M. (2010). Evaluating the effects of ideology on public understanding of climate change science: How to improve communication across ideological divides? *Public Understanding of Climate Change*, 19(6), 743–761.

INDEX

© The Author(s) 2017
A. Corner, J. Clarke, *Talking Climate*,
DOI 10.1007/978-3-319-46744-3